Audio Engineering

Audio Engineering

Editor: John McGee

NY RESEARCH
P R E S S

New York

Published by NY Research Press
118-35 Queens Blvd., Suite 400,
Forest Hills, NY 11375, USA
www.nyresearchpress.com

Audio Engineering
Edited by John McGee

International Standard Book Number: 978-1-63238-590-1 (Hardback)

Cataloging-in-Publication Data

Audio engineering / edited by John McGee.
    p. cm.
Includes bibliographical references and index.
ISBN 978-1-63238-590-1
1. Acoustical engineering. 2. Sound--Recording and reproducing.
3. Sound--Recording and reproducing--Digital techniques. I. McGee, John.
TA365 .A93 2018
620.2--dc23

Contents

Permissions

Index

Preface

Audio engineering refers to the study and research related to the manipulation of sound using elements like mixing, equalization, reinforcement and electronic effects. It includes the study of music and also deals with the development of audio technology. This field has many sub-disciplines like musical acoustics, audio-signal processing, electroacoustics, speech, architectural acoustics, psychoacoustics, etc. This book is a valuable compilation of topics, ranging from the basic to the most complex theories and principles in the field of audio engineering. Some of the diverse topics covered in it address the varied branches that fall under this category. This textbook is a complete source of knowledge on the present status of this important field.

A short introduction to every chapter is written below to provide an overview of the content of the book:

Chapter 1 - Sound is a vibration that travels through a medium like air or water. The speed of sound depends on the medium it is traveling through. The important aspects related to the phenomenon of sound are soundscape, timbre, loudness, etc. This chapter will provide an integrated understanding of sound; **Chapter 2 -** Audio frequency is the frequency that is sensed by the human ear. The unit used for audio frequency is hertz. This chapter has been carefully written to provide an easy understanding of the varied facets of audio frequency; **Chapter 3 -** Devices that can modify, create and convert audio signals into an output are known as audio electronics. The other equipments discussed are analog-to-digital converter, loudspeaker, mixing console, tape recorder and music sequencer. The topics discussed in the chapter are of great importance to broaden the existing knowledge on the equipments used in audio engineering; **Chapter 4 -** The act of mixing several tracks or recordings is known as audio mixing. The techniques used in the process mainly depend upon the quality and particular genres the recordings belong to. The chapter serves as a source to understand the major categories related to audio mixing; **Chapter 5 -** Amplifiers are devices used to increase the voltage or power of a signal. Headphone amplifier, class-t amplifier, carbon microphone, and valve audio amplifier are some examples of audio amplifiers that have been listed in this chapter. The section on audio amplifiers offers an insightful focus, keeping in mind the complex subject matter.

I extend my sincere thanks to the publisher for considering me worthy of this task. Finally, I thank my family for being a source of support and help.

Editor

An Introduction to Sound

Sound is a vibration that travels through a medium like air or water. The speed of sound depends on the medium it is traveling through. The important aspects related to the phenomenon of sound are soundscape, timbre, loudness, etc. This chapter will provide an integrated understanding of sound.

Sound

A drum produces sound via a vibrating membrane.

In physics, sound is a vibration that propagates as a typically audible mechanical wave of pressure and displacement, through a transmission medium such as air or water. In physiology and psychology, sound is the *reception* of such waves and their *perception* by the brain. Humans can hear sound waves with frequencies between about 20 Hz and 20 kHz. Sound above 20 kHz is ultrasound and below 20 Hz is infrasound. Other animals have different hearing ranges.

Acoustics

Acoustics is the interdisciplinary science that deals with the study of mechanical waves in gases, liquids, and solids including vibration, sound, ultrasound, and infrasound. A scientist who works in the field of acoustics is an *acoustician*, while someone working in the field of acoustical engineering may be called an *acoustical engineer*. An audio engineer, on the other hand, is concerned with the recording, manipulation, mixing, and reproduction of sound.

Applications of acoustics are found in almost all aspects of modern society, subdisciplines include aeroacoustics, audio signal processing, architectural acoustics, bioacoustics, electro-acoustics, environmental noise, musical acoustics, noise control, psychoacoustics, speech, ultrasound, underwater acoustics, and vibration.

Definition

Sound is defined by ANSI/ASA S1.1-2013 as "(a) Oscillation in pressure, stress, particle displacement, particle velocity, etc., propagated in a medium with internal forces (e.g., elastic or viscous), or the superposition of such propagated oscillation. (b) Auditory sensation evoked by the oscillation described in (a)."

Physics of Sound

Sound can propagate through a medium such as air, water and solids as longitudinal waves and also as a transverse wave in solids. The sound waves are generated by a sound source, such as the vibrating diaphragm of a stereo speaker. The sound source creates vibrations in the surrounding medium. As the source continues to vibrate the medium, the vibrations propagate away from the source at the speed of sound, thus forming the sound wave. At a fixed distance from the source, the pressure, velocity, and displacement of the medium vary in time. At an instant in time, the pressure, velocity, and displacement vary in space. Note that the particles of the medium do not travel with the sound wave. This is intuitively obvious for a solid, and the same is true for liquids and gases (that is, the vibrations of particles in the gas or liquid transport the vibrations, while the *average* position of the particles over time does not change). During propagation, waves can be reflected, refracted, or attenuated by the medium.

Spherical compression (longitudinal) waves

The behavior of sound propagation is generally affected by three things:

- A complex relationship between the density and pressure of the medium. This relationship, affected by temperature, determines the speed of sound within the medium.

- Motion of the medium itself. If the medium is moving, this movement may increase or decrease the absolute speed of the sound wave depending on the direction of the movement. For example, sound moving through wind will have its speed of propagation increased by the speed of the wind if the sound and wind are moving in the same direction. If the sound and wind are moving in opposite directions, the speed of the sound wave will be decreased by the speed of the wind.

- The viscosity of the medium. Medium viscosity determines the rate at which sound is attenuated. For many media, such as air or water, attenuation due to viscosity is negligible.

When sound is moving through a medium that does not have constant physical properties, it may be refracted (either dispersed or focused).

The mechanical vibrations that can be interpreted as sound can travel through all forms of matter: gases, liquids, solids, and plasmas. The matter that supports the sound is called the medium. Sound cannot travel through a vacuum.

Longitudinal and Transverse Waves

Sound is transmitted through gases, plasma, and liquids as longitudinal waves, also called compression waves. It requires a medium to propagate. Through solids, however, it can be transmitted as both longitudinal waves and transverse waves. Longitudinal sound waves are waves of alternating pressure deviations from the equilibrium pressure, causing local regions of compression and rarefaction, while transverse waves (in solids) are waves of alternating shear stress at right angle to the direction of propagation.

Sound waves may be "viewed" using parabolic mirrors and objects that produce sound.

The energy carried by an oscillating sound wave converts back and forth between the potential energy of the extra compression (in case of longitudinal waves) or lateral displacement strain (in case of transverse waves) of the matter, and the kinetic energy of the displacement velocity of particles of the medium.

Sound Wave Properties and Characteristics

Although there are many complexities relating to the transmission of sounds, at the point of reception (i.e. the ears), sound is readily dividable into two simple elements: pressure and time. These fundamental elements form the basis of all sound waves. They can be used to describe, in absolute terms, every sound we hear.

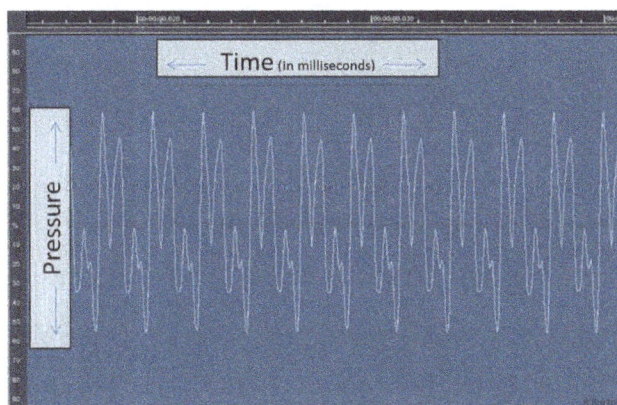

A 'pressure over time' graph of a 20 ms recording of a clarinet tone demonstrates the two fundamental elements of sound: Pressure and Time.

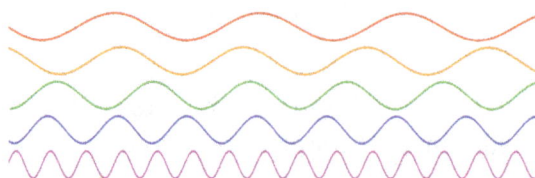

Sounds can be represented as a mixture of their component Sinusoidal waves of different frequencies.
The bottom waves have higher frequencies than those above. The horizontal axis represents time.

However, in order to understand the sound more fully, a complex wave such as this is usually separated into its component parts, which are a combination of various sound wave frequencies (and noise).

Sound waves are often simplified to a description in terms of sinusoidal plane waves, which are characterized by these generic properties:

- Frequency, or its inverse, wavelength
- Amplitude, sound pressure or Intensity
- Speed of sound
- Direction

Sound that is perceptible by humans has frequencies from about 20 Hz to 20,000 Hz. In air at standard temperature and pressure, the corresponding wavelengths of sound waves range from 17 m to 17 mm. Sometimes speed and direction are combined as a velocity vector; wave number and direction are combined as a wave vector.

Transverse waves, also known as shear waves, have the additional property, *polarization*, and are not a characteristic of sound waves.

Speed of Sound

U.S. Navy F/A-18 approaching the sound barrier. The white halo is formed by condensed water droplets thought to result from a drop in air pressure around the aircraft.

The speed of sound depends on the medium that the waves pass through, and is a fundamental property of the material. The first significant effort towards the measure of the speed of sound was made by Newton. He believed that the speed of sound in a particular substance was equal to the square root of the pressure acting on it divided by its density:

$$c = \sqrt{\frac{p}{\rho}}$$

This was later proven wrong when found to incorrectly derive the speed. The French mathematician Laplace corrected the formula by deducing that the phenomenon of sound travelling is not isothermal, as believed by Newton, but adiabatic. He added another factor to the equation—$gamma$—and multiplied $\sqrt{\gamma}$ by $\sqrt{\frac{p}{\rho}}$, thus coming up with the equation $c = \sqrt{\gamma \cdot \frac{p}{\rho}}$. Since $K = \gamma \cdot p$, the final equation came up to be $c = \sqrt{\frac{K}{\rho}}$, which is also known as the Newton-Laplace equation. In this equation, K = elastic bulk modulus, c = velocity of sound, and ρ = density. Thus, the speed of sound is proportional to the square root of the ratio of the bulk modulus of the medium to its density.

Those physical properties and the speed of sound change with ambient conditions. For example, the speed of sound in gases depends on temperature. In 20 °C (68 °F) air at sea level, the speed of sound is approximately 343 m/s (1,230 km/h; 767 mph) using the formula "$v = (331 + 0.6\ T)$ m/s". In fresh water, also at 20 °C, the speed of sound is approximately 1,482 m/s (5,335 km/h; 3,315 mph). In steel, the speed of sound is about 5,960 m/s (21,460 km/h; 13,330 mph). The speed of sound is also slightly sensitive, being subject to a second-order anharmonic effect, to the sound amplitude, which means that there are non-linear propagation effects, such as the production of harmonics and mixed tones not present in the original sound.

Perception of Sound

A distinct use of the term *sound* from its use in physics is that in physiology and psychology, where the term refers to the subject of *perception* by the brain. The field of psychoacoustics is dedicated to such studies. Historically the word "sound" referred exclusively to an effect in the mind. Webster's 1947 dictionary defined sound as: "that which is heard; the effect which is produced by the vibration of a body affecting the ear." This meant (at least in 1947) the correct response to the question: "if a tree falls in the forest with no one to hear it fall, does it make a sound?" was "no". However, owing to contemporary usage, definitions of sound as a physical effect are prevalent in most dictionaries. Consequently, the answer to the same question is now "yes, a tree falling in the forest with no one to hear it fall does make a sound".

The physical reception of sound in any hearing organism is limited to a range of frequencies. Humans normally hear sound frequencies between approximately 20 Hz and 20,000 Hz (20 kHz), The upper limit decreases with age. Sometimes *sound* refers to only those vibrations with frequencies that are within the hearing range for humans or sometimes it relates to a particular animal. Other species have different ranges of hearing. For example, dogs can perceive vibrations higher than 20 kHz, but are deaf below 40 Hz.

As a signal perceived by one of the major senses, sound is used by many species for detecting danger, navigation, predation, and communication. Earth's atmosphere, water, and virtually any physical phenomenon, such as fire, rain, wind, surf, or earthquake, produces (and is characterized by) its unique sounds. Many species, such as frogs, birds, marine and terrestrial mammals, have also developed special organs to produce sound. In some species, these produce song and speech.

Furthermore, humans have developed culture and technology (such as music, telephone and radio) that allows them to generate, record, transmit, and broadcast sound.

Elements of Sound Perception

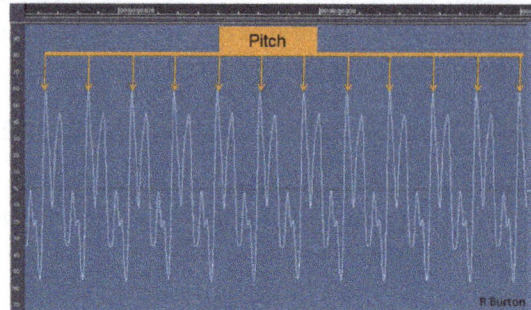

(a) Pitch perception

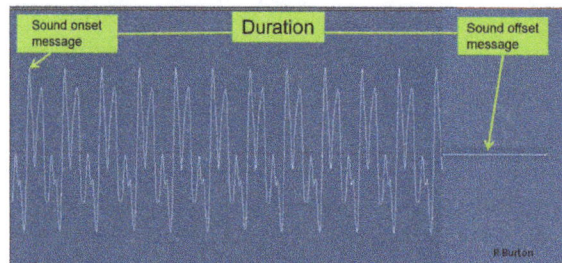

(b) Duration perception

There are six experimentally separable ways in which sound waves are analysed. They are: pitch, duration, loudness, timbre, sonic texture and spatial location.

Pitch

Pitch is perceived as how "low" or "high" a sound is and represents the cyclic, repetitive nature of the vibrations that make up sound. For simple sounds, pitch relates to the frequency of the slowest vibration in the sound (called the fundamental harmonic). In the case of complex sounds, pitch perception can vary. Sometimes individuals identify different pitches for the same sound, based on their personal experience of particular sound patterns. Selection of a particular pitch is determined by pre-conscious examination of vibrations, including their frequencies and the balance between them. Specific attention is given to recognising potential harmonics. Every sound is placed on a pitch continuum from low to high. For example: white noise (random noise spread evenly across all frequencies) sounds higher in pitch than pink noise (random noise spread evenly across octaves) as white noise has more high frequency content. Figure 1 shows an example of pitch recognition. During the listening process, each sound is analysed for a repeating pattern and the results forwarded to the auditory cortex as a single pitch of a certain height (octave) and chroma (note name).

Duration

Duration is perceived as how "long" or "short" a sound is and relates to onset and offset signals created by nerve responses to sounds. The duration of a sound usually lasts from the time the sound is first noticed until the sound is identified as having changed or ceased. Sometimes this is not directly related to the physical duration of a sound. For example; in a noisy environment, gapped sounds (sounds that stop and start) can sound as if they are continuous because the offset messages are missed owing to disruptions from noises in the same general bandwidth. This can be of great benefit in understanding distorted messages such as radio signals that suffer from interference, as (owing to this effect) the message is heard as if it was continuous. Figure b gives an example of duration identification. When a new sound is noticed, a sound onset message is sent to the auditory cortex. When the repeating pattern is missed, a sound offset messages is sent.

Loudness

Loudness is perceived as how "loud" or "soft" a sound is and relates to the totalled number of auditory nerve stimulations over short cyclic time periods, most likely over the duration of theta wave cycles. This means that at short durations, a very short sound can sound softer than a longer sound even though they are presented at the same intensity level. Past around 200 ms this is no longer the case and the duration of the sound no longer affects the apparent loudness of the sound. Loudness information is summed over a period of about 200 ms before being sent to the auditory cortex. Louder signals create a greater 'push' on the Basilar membrane and thus stimulate more nerves, creating a stronger loudness signal. A more complex signal also creates more nerve firings and so sounds louder (for the same wave amplitude) than a simpler sound, such as a sine wave.

Timbre

Timbre is perceived as the quality of different sounds (e.g. the thud of a fallen rock, the whir of a drill, the tone of a musical instrument or the quality of a voice) and represents the pre-conscious allocation of a sonic identity to a sound (e.g. "it's an oboe!"). This identity is based on information gained from frequency transients, noisiness, unsteadiness, perceived pitch and the spread and intensity of overtones in the sound over an extended time frame. The way a sound changes over time provides most of the information for timbre identification. Even though a small section of the wave form from each instrument looks very similar, differences in changes over time between the clarinet and the piano are evident in both loudness and harmonic content. Less noticeable are the different noises heard, such as air hisses for the clarinet and hammer strikes for the piano.

Sonic Texture

Sonic texture relates to the number of sound sources and the interaction between them. The word 'texture', in this context, relates to the cognitive separation of auditory objects. In music, texture is often referred to as the difference between unison, polyphony and homophony, but it can also relate (for example) to a busy cafe; a sound which might be referred to as 'cacophony'. However texture refers to more than this. The texture of an orchestral piece is very different to the texture of a brass quartet because of the different numbers of players. The texture of a market place is very different to a school hall because of the differences in the various sound sources.

Spatial Location

Spatial location represents the cognitive placement of a sound in an environmental context; including the placement of a sound on both the horizontal and vertical plane, the distance from the sound source and the characteristics of the sonic environment. In a thick texture, it is possible to identify multiple sound sources using a combination of spatial location and timbre identification. It is the main reason why we can pick the sound of an oboe in an orchestra and the words of a single person at a cocktail party.

Noise

Noise is a term often used to refer to an unwanted sound. In science and engineering, noise is an undesirable component that obscures a wanted signal. However, in sound perception it can often be used to identify the source of a sound and is an important component of timbre perception.

Soundscape

Soundscape is the component of the acoustic environment that can be perceived by humans. The acoustic environment is the combination of all sounds (whether audible to humans or not) within a given area as modified by the environment and understood by people, in context of the surrounding environment.

Sound Pressure Level

Sound pressure is the difference, in a given medium, between average local pressure and the pressure in the sound wave. A square of this difference (i.e., a square of the deviation from the equilibrium pressure) is usually averaged over time and/or space, and a square root of this average provides a root mean square (RMS) value. For example, 1 Pa RMS sound pressure (94 dBSPL) in atmospheric air implies that the actual pressure in the sound wave oscillates between (1 atm $-\sqrt{2}$ Pa) and (1 atm $+\sqrt{2}$ Pa), that is between 101323.6 and 101326.4 Pa. As the human ear can detect sounds with a wide range of amplitudes, sound pressure is often measured as a level on a logarithmic decibel scale. The sound pressure level (SPL) or L_p is defined as

$$L_p = 10\log_{10}\left(\frac{p^2}{p_{ref}^2}\right) = 20\log_{10}\left(\frac{p}{p_{ref}}\right) \text{ dB}$$

where p is the root-mean-square sound pressure and p_{ref} is a reference sound pressure. Commonly used reference sound pressures, defined in the standard ANSI S1.1-1994, are 20 µPa in air and 1 µPa in water. Without a specified reference sound pressure, a value expressed in decibels cannot represent a sound pressure level.

Since the human ear does not have a flat spectral response, sound pressures are often frequency weighted so that the measured level matches perceived levels more closely. The International Electrotechnical Commission (IEC) has defined several weighting schemes. A-weighting attempts to match the response of the human ear to noise and A-weighted sound pressure levels are labeled dBA. C-weighting is used to measure peak levels.

Speed of Sound

The speed of sound is the distance travelled per unit time by a sound wave as it propagates through an elastic medium. In dry air at 20° C (68° F), the speed of sound is 343 metres per second (1,125 ft/s; 1,235 km/h; 767 mph; 667 km), or a kilometre in 2.91 s or a mile in 4.69 s.

The speed of sound in an ideal gas depends only on its temperature and composition. The speed has a weak dependence on frequency and pressure in ordinary air, deviating slightly from ideal behavior.

In common everyday speech, *speed of sound* refers to the speed of sound waves in air. However, the speed of sound varies from substance to substance: sound travels most slowly in gases; it travels faster in liquids; and faster still in solids. For example, (as noted above), sound travels at 343 m/s in air; it travels at 1,484 m/s in water (4.3 times as fast as in air); and at 5,120 m/s in iron. In an exceptionally stiff material such as diamond, sound travels at 12,000 m/s; which is around the maximum speed that sound will travel under normal conditions.

Sound waves in solids are composed of compression waves (just as in gases and liquids), and a different type of sound wave called a shear wave, which occurs only in solids. Shear waves in solids usually travel at different speeds, as exhibited in seismology. The speed of compression waves in solids is determined by the medium's compressibility, shear modulus and density. The speed of shear waves is determined only by the solid material's shear modulus and density.

In fluid dynamics, the speed of sound in a fluid medium (gas or liquid) is used as a relative measure for the speed of an object moving through the medium. The ratio of the speed of an object to the speed of sound in the fluid is called the object's Mach number. Objects moving at speeds greater than *Mach1* are said to be traveling at supersonic speeds.

History

Sir Isaac Newton computed the speed of sound in air as 979 feet per second (298 m/s), which is too low by about 15%, but had neglected the effect of fluctuating temperature; that was later rectified by Laplace.

During the 17th century, there were several attempts to measure the speed of sound accurately, including attempts by Marin Mersenne in 1630 (1,380 Parisian feet per second), Pierre Gassendi in 1635 (1,473 Parisian feet per second) and Robert Boyle (1,125 Parisian feet per second).

In 1709, the Reverend William Derham, Rector of Upminster, published a more accurate measure of the speed of sound, at 1,072 Parisian feet per second. Derham used a telescope from the tower of the church of St Laurence, Upminster to observe the flash of a distant shotgun being fired, and then measured the time until he heard the gunshot with a half second pendulum. Measurements were made of gunshots from a number of local landmarks, including North Ockendon church. The distance was known by triangulation, and thus the speed that the sound had travelled was calculated.

Basic Concept

The transmission of sound can be illustrated by using a model consisting of an array of balls interconnected by springs. For real material the balls represent molecules and the springs represent the bonds between them. Sound passes through the model by compressing and expanding the springs, transmitting energy to neighbouring balls, which transmit energy to their springs, and so on. The speed of sound through the model depends on the stiffness of the springs, and the mass of the balls. As long as the spacing of the balls remains constant, stiffer springs transmit energy more quickly, and more massive balls transmit energy more slowly. Effects like dispersion and reflection can also be understood using this model.

In a real material, the stiffness of the springs is called the elastic modulus, and the mass corresponds to the density. All other things being equal (ceteris paribus), sound will travel more slowly in spongy materials, and faster in stiffer ones. For instance, sound will travel 1.59 times faster in nickel than in bronze, due to the greater stiffness of nickel at about the same density. Similarly, sound travels about 1.41 times faster in light hydrogen (protium) gas than in heavy hydrogen (deuterium) gas, since deuterium has similar properties but twice the density. At the same time, "compression-type" sound will travel faster in solids than in liquids, and faster in liquids than in gases, because the solids are more difficult to compress than liquids, while liquids in turn are more difficult to compress than gases.

Some textbooks mistakenly state that the speed of sound increases with increasing density. This is usually illustrated by presenting data for three materials, such as air, water and steel, which also have vastly different compressibilities which more than make up for the density differences. An illustrative example of the two effects is that sound travels only 4.3 times faster in water than air, despite enormous differences in compressibility of the two media. The reason is that the larger density of water, which works to *slow* sound in water relative to air, nearly makes up for the compressibility differences in the two media.

Compression and Shear Waves

In a gas or liquid, sound consists of compression waves. In solids, waves propagate as two different types. A longitudinal wave is associated with compression and decompression in the direction of travel, and is the same process in gases and liquids, with an analogous compression-type wave in solids. Only compression waves are supported in gases and liquids. An additional type of wave, the transverse wave, also called a shear wave, occurs only in solids because only solids support elastic deformations. It is due to elastic deformation of the medium perpendicular to the direction of wave travel; the direction of shear-deformation is called the "polarization" of this type of wave. In general, transverse waves occur as a pair of orthogonal polarizations.

These different waves (compression waves and the different polarizations of shear waves) may have different speeds at the same frequency. Therefore, they arrive at an observer at different times, an extreme example being an earthquake, where sharp compression waves arrive first, and rocking transverse waves seconds later.

The speed of a compression wave in fluid is determined by the medium's compressibility and density. In solids, the compression waves are analogous to those in fluids, depending on compressibility, density, and the additional factor of shear modulus. The speed of shear waves, which can occur only in solids, is determined simply by the solid material's shear modulus and density.

Equations

The speed of sound in mathematical notation is conventionally represented by c, from the Latin *celeritas* meaning "velocity".

In general, the speed of sound c is given by the Newton–Laplace equation:

$$c = \sqrt{\frac{K_s}{\rho}},$$

where

- K_s is a coefficient of stiffness, the isentropic bulk modulus (or the modulus of bulk elasticity for gases);

- ρ is the density.

Thus the speed of sound increases with the stiffness (the resistance of an elastic body to deformation by an applied force) of the material, and decreases with increase in density. For ideal gases the bulk modulus K is simply the gas pressure multiplied by the dimensionless adiabatic index, which is about 1.4 for air under normal conditions of pressure and temperature.

For general equations of state, if classical mechanics is used, the speed of sound c is given by

$$c = \sqrt{\left(\frac{\partial p}{\partial \rho}\right)_s},$$

where

- p is the pressure;

- ρ is the density and the derivative is taken isentropically, that is, at constant entropy s.

If relativistic effects are important, the speed of sound is calculated from the relativistic Euler equations.

In a non-dispersive medium, the speed of sound is independent of sound frequency, so the speeds of energy transport and sound propagation are the same for all frequencies. Air, a mixture of oxygen and nitrogen, constitutes a non-dispersive medium. However, air does contain a small amount of CO_2 which *is* a dispersive medium, and causes dispersion to air at ultrasonic frequencies (> 28 kHz).

In a dispersive medium, the speed of sound is a function of sound frequency, through the dispersion relation. Each frequency component propagates at its own speed, called the phase velocity, while the energy of the disturbance propagates at the group velocity. The same phenomenon occurs with light waves; see optical dispersion for a description.

Dependence on the Properties of the Medium

The speed of sound is variable and depends on the properties of the substance through which the wave is travelling. In solids, the speed of transverse (or shear) waves depends on the shear deformation under shear stress (called the shear modulus), and the density of the medium. Longitudinal (or compression) waves in solids depend on the same two factors with the addition of a dependence on compressibility.

In fluids, only the medium's compressibility and density are the important factors, since fluids do not transmit shear stresses. In heterogeneous fluids, such as a liquid filled with gas bubbles, the density of the liquid and the compressibility of the gas affect the speed of sound in an additive manner, as demonstrated in the hot chocolate effect.

In gases, adiabatic compressibility is directly related to pressure through the heat capacity ratio (adiabatic index), while pressure and density are inversely related to the temperature and molecular weight, thus making only the completely independent properties of *temperature and molecular structure* important (heat capacity ratio may be determined by temperature and molecular structure, but simple molecular weight is not sufficient to determine it).

In low molecular weight gases such as helium, sound propagates faster as compared to heavier gases such as xenon. For monatomic gases, the speed of sound is about 75% of the mean speed that the atoms move in that gas.

For a given ideal gas the molecular composition is fixed, and thus the speed of sound depends only on its temperature. At a constant temperature, the gas pressure has no effect on the speed of sound, since the density will increase, and since pressure and density (also proportional to pressure) have equal but opposite effects on the speed of sound, and the two contributions cancel out exactly. In a similar way, compression waves in solids depend both on compressibility and density—just as in liquids—but in gases the density contributes to the compressibility in such a way that some part of each attribute factors out, leaving only a dependence on temperature, molecular weight, and heat capacity ratio which can be independently derived from temperature and molecular composition. Thus, for a single given gas (assuming the molecular weight does not change) and over a small temperature range (for which the heat capacity is relatively constant), the speed of sound becomes dependent on only the temperature of the gas.

In non-ideal gas behavior regimen, for which the van der Waals gas equation would be used, the proportionality is not exact, and there is a slight dependence of sound velocity on the gas pressure.

Humidity has a small but measurable effect on the speed of sound (causing it to increase by about 0.1%–0.6%), because oxygen and nitrogen molecules of the air are replaced by lighter molecules of water. This is a simple mixing effect.

Altitude Variation and Implications for Atmospheric Acoustics

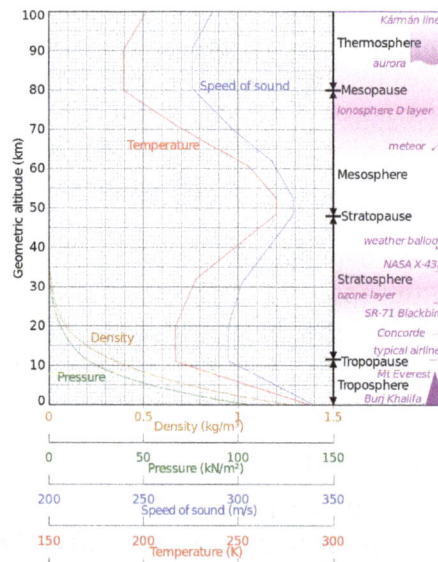

Density and pressure decrease smoothly with altitude, but temperature (red) does not. The speed of sound (blue) depends only on the complicated temperature variation at altitude and can be calculated from it, since isolated density and pressure effects on the speed of sound cancel each other. Speed of sound increases with height in two regions of the stratosphere and thermosphere, due to heating effects in these regions.

In the Earth's atmosphere, the chief factor affecting the speed of sound is the temperature. For a given ideal gas with constant heat capacity and composition, the speed of sound is dependent *solely* upon temperature; see Details below. In such an ideal case, the effects of decreased density and decreased pressure of altitude cancel each other out, save for the residual effect of temperature.

Since temperature (and thus the speed of sound) decreases with increasing altitude up to 11 km, sound is refracted upward, away from listeners on the ground, creating an acoustic shadow at some distance from the source. The decrease of the speed of sound with height is referred to as a negative sound speed gradient.

However, there are variations in this trend above 11 km. In particular, in the stratosphere above about 20 km, the speed of sound increases with height, due to an increase in temperature from heating within the ozone layer. This produces a positive speed of sound gradient in this region. Still another region of positive gradient occurs at very high altitudes, in the aptly-named thermosphere above 90 km.

Practical Formula for Dry Air

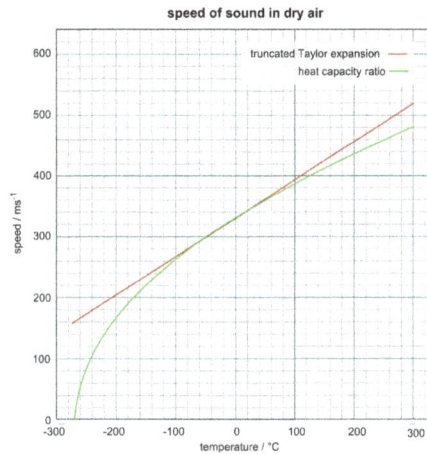

Approximation of the speed of sound in dry air based on the heat capacity ratio (in green) against the truncated Taylor expansion (in red).

The approximate speed of sound in dry (0% humidity) air, in meters per second, at temperatures near 0° C, can be calculated from

$$c_{air} = (331.3 + 0.606 \cdot \vartheta) \text{ m/s,}$$

where ϑ is the temperature in degrees Celsius (°C).

This equation is derived from the first two terms of the Taylor expansion of the following more accurate equation:

$$c_{air} = 331.3 \sqrt{1 + \frac{\vartheta}{273.15}} \quad \text{m/s.}$$

Dividing the first part, and multiplying the second part, on the right hand side, by $\sqrt{273.15}$ gives the exactly equivalent form

$$c_{air} = 20.05\sqrt{\vartheta + 273.15} \quad m/s.$$

The value of 331.3 m/s, which represents the speed at 0 °C (or 273.15 K), is based on theoretical (and some measured) values of the heat capacity ratio, γ, as well as on the fact that at 1 atm real air is very well described by the ideal gas approximation. Commonly found values for the speed of sound at 0 °C may vary from 331.2 to 331.6 due to the assumptions made when it is calculated. If ideal gas γ is assumed to be 7/5 = 1.4 exactly, the 0 °C speed is calculated to be 331.3 m/s, the coefficient used above.

This equation is correct to a much wider temperature range, but still depends on the approximation of heat capacity ratio being independent of temperature, and for this reason will fail, particularly at higher temperatures. It gives good predictions in relatively dry, cold, low pressure conditions, such as the Earth's stratosphere. The equation fails at extremely low pressures and short wavelengths, due to dependence on the assumption that the wavelength of the sound in the gas is much longer than the average mean free path between gas molecule collisions.

A graph comparing results of the two equations is at right, using the slightly different value of 331.5 m/s for the speed of sound at 0 °C.

Details

Speed of Sound in Ideal Gases and Air

For an ideal gas, K (the bulk modulus in equations above, equivalent to C, the coefficient of stiffness in solids) is given by

$$K = \gamma \cdot p,$$

thus, from the Newton–Laplace equation above, the speed of sound in an ideal gas is given by

$$c = \sqrt{\gamma \cdot \frac{p}{\rho}},$$

where

- γ is the adiabatic index also known as the *isentropic expansion factor*. It is the ratio of specific heats of a gas at a constant-pressure to a gas at a constant-volume(C_p / C_v), and arises because a classical sound wave induces an adiabatic compression, in which the heat of the compression does not have enough time to escape the pressure pulse, and thus contributes to the pressure induced by the compression;

- p is the pressure;

- ρ is the density.

Using the ideal gas law to replace p with nRT/V, and replacing ρ with nM/V, the equation for an ideal gas becomes

$$c_{ideal} = \sqrt{\gamma \cdot \frac{p}{\rho}} = \sqrt{\frac{\gamma \cdot R \cdot T}{M}} = \sqrt{\frac{\gamma \cdot k \cdot T}{m}},$$

where

- c_{ideal} is the speed of sound in an ideal gas;
- R (approximately $8.314{,}5\ \text{J} \cdot \text{mol}^{-1} \cdot \text{K}^{-1}$) is the molar gas constant(universal gas constant);
- k is the Boltzmann constant;
- γ (gamma) is the adiabatic index. At room temperature, where thermal energy is fully partitioned into rotation (rotations are fully excited) but quantum effects prevent excitation of vibrational modes, the value is $7/5 = 1.400$ for diatomic molecules, according to kinetic theory. Gamma is actually experimentally measured over a range from $1.399{,}1$ to 1.403 at 0 °C, for air. Gamma is exactly $5/3 = 1.6667$ for monatomic gases such as noble gases;
- T is the absolute temperature;
- M is the molar mass of the gas. The mean molar mass for dry air is about $0.028{,}964{,}5$ kg/mol;
- n is the number of moles;
- m is the mass of a single molecule.

This equation applies only when the sound wave is a small perturbation on the ambient condition, and the certain other noted conditions are fulfilled, as noted below. Calculated values for c_{air} have been found to vary slightly from experimentally determined values.

Newton famously considered the speed of sound before most of the development of thermodynamics and so incorrectly used isothermal calculations instead of adiabatic. His result was missing the factor of γ but was otherwise correct.

Numerical substitution of the above values gives the ideal gas approximation of sound velocity for gases, which is accurate at relatively low gas pressures and densities (for air, this includes standard Earth sea-level conditions). Also, for diatomic gases the use of $\gamma = 1.400{,}0$ requires that the gas exists in a temperature range high enough that rotational heat capacity is fully excited (i.e., molecular rotation is fully used as a heat energy "partition" or reservoir); but at the same time the temperature must be low enough that molecular vibrational modes contribute no heat capacity (i.e., insignificant heat goes into vibration, as all vibrational quantum modes above the minimum-energy-mode, have energies too high to be populated by a significant number of molecules at this temperature). For air, these conditions are fulfilled at room temperature, and also temperatures considerably below room temperature.

For air, we use a simplified symbol

$$R_* = R / M_{\text{air}}.$$

Additionally, if temperatures in degrees Celsius(°C) are to be used to calculate air speed in the region near 273 kelvin, then Celsius temperature $\theta = T - 273.15$ may be used. Then

$$c_{\text{ideal}} = \sqrt{\gamma \cdot R_* \cdot T} = \sqrt{\gamma \cdot R_* \cdot (\vartheta + 273.15)},$$

$$c_{ideal} = \sqrt{\gamma \cdot R_* \cdot 273.15} \cdot \sqrt{1 + \frac{\vartheta}{273.15}}.$$

For dry air, where θ (theta) is the temperature in degrees Celsius(°C).

Making the following numerical substitutions,

$$R = 8.314,510 \text{ J}/(\text{mol} \cdot \text{K})$$

is the molar gas constant in J/mole/Kelvin, and

$$M_{air} = 0.028,964,5 \text{ kg/mol}$$

is the mean molar mass of air, in kg; and using the ideal diatomic gas value of $\gamma = 1.4000$.

Then

$$c_{air} = 331.3 \sqrt{1 + \frac{\vartheta°C}{273.15°C}} \quad \text{m/s.}$$

Using the first two terms of the Taylor expansion:

$$c_{air} = 331.3 \left(1 + \frac{\vartheta°C}{2 \cdot 273.15°C}\right) \quad \text{m/s,}$$

$$c_{air} = (331.3 + 0.606°C^{-1} \cdot \vartheta) \quad \text{m/s.}$$

The derivation includes the first two equations given in the *Practical formula for dry* air section above.

Effects Due to Wind Shear

The speed of sound varies with temperature. Since temperature and sound velocity normally decrease with increasing altitude, sound is refracted upward, away from listeners on the ground, creating an acoustic shadow at some distance from the source. Wind shear of 4 m/(s · km) can produce refraction equal to a typical temperature lapse rate of 7.5 °C/km. Higher values of wind gradient will refract sound downward toward the surface in the downwind direction, eliminating the acoustic shadow on the downwind side. This will increase the audibility of sounds downwind. This downwind refraction effect occurs because there is a wind gradient; the sound is not being carried along by the wind.

For sound propagation, the exponential variation of wind speed with height can be defined as follows:

$$U(h) = U(0)h^{\varsigma},$$

$$\frac{dU}{dH}(h) = \varsigma \frac{U(h)}{h},$$

where

- $U(h)$ is the speed of the wind at height h;
- ζ is the exponential coefficient based on ground surface roughness, typically between 0.08 and 0.52;
- $dU/dH(h)$ is the expected wind gradient at height h.

In the 1862 American Civil War Battle of Iuka, an acoustic shadow, believed to have been enhanced by a northeast wind, kept two divisions of Union soldiers out of the battle, because they could not hear the sounds of battle only 10 km (six miles) downwind.

Tables

In the standard atmosphere:

- T_0 is 273.15 K (= 0 °C = 32 °F), giving a theoretical value of 331.3 m/s (= 1086.9 ft/s = 1193 km/h = 741.1 mph = 644.0 kn). Values ranging from 331.3-331.6 may be found in reference literature, however;
- T_{20} is 293.15 K (= 20 °C = 68 °F), giving a value of 343.2 m/s (= 1126.0 ft/s = 1236 km/h = 767.8 mph = 667.2 kn);
- T_{25} is 298.15 K (= 25 °C = 77 °F), giving a value of 346.1 m/s (= 1135.6 ft/s = 1246 km/h = 774.3 mph = 672.8 kn).

In fact, assuming an ideal gas, the speed of sound c depends on temperature only, not on the pressure or density (since these change in lockstep for a given temperature and cancel out). Air is almost an ideal gas. The temperature of the air varies with altitude, giving the following variations in the speed of sound using the standard atmosphere—*actual conditions may vary.*

Effect of temperature on properties of air			
Temperature T (°C)	Speed of sound c (m/s)	Density of air ρ (kg/m³)	Characteristic specific acoustic impedance z_0 (Pa·s/m)
35	351.88	1.1455	403.2
30	349.02	1.1644	406.5
25	346.13	1.1839	409.4
20	343.21	1.2041	413.3
15	340.27	1.2250	416.9
10	337.31	1.2466	420.5
5	334.32	1.2690	424.3
0	331.30	1.2922	428.0
−5	328.25	1.3163	432.1
−10	325.18	1.3413	436.1
−15	322.07	1.3673	440.3
−20	318.94	1.3943	444.6
−25	315.77	1.4224	449.1

Given normal atmospheric conditions, the temperature, and thus speed of sound, varies with altitude:

Altitude	Temperature	m/s	km/h	mph	kn
Sea level	15 °C (59 °F)	340	1,225	761	661
11,000 m–20,000 m (Cruising altitude of commercial jets, and first supersonic flight)	−57 °C (−70 °F)	295	1,062	660	573
29,000 m (Flight of X-43A)	−48 °C (−53 °F)	301	1,083	673	585

Effect of Frequency and Gas Composition

General Physical Considerations

The medium in which a sound wave is travelling does not always respond adiabatically, and as a result the speed of sound can vary with frequency.

The limitations of the concept of speed of sound due to extreme attenuation are also of concern. The attenuation which exists at sea level for high frequencies applies to successively lower frequencies as atmospheric pressure decreases, or as the mean free path increases. For this reason, the concept of speed of sound (except for frequencies approaching zero) progressively loses its range of applicability at high altitudes. The standard equations for the speed of sound apply with reasonable accuracy only to situations in which the wavelength of the soundwave is considerably longer than the mean free path of molecules in a gas.

The molecular composition of the gas contributes both as the mass (M) of the molecules, and their heat capacities, and so both have an influence on speed of sound. In general, at the same molecular mass, monatomic gases have slightly higher speed of sound (over 9% higher) because they have a higher γ (5/3 = 1.66...) than diatomics do (7/5 = 1.4). Thus, at the same molecular mass, the speed of sound of a monatomic gas goes up by a factor of

$$\frac{c_{gas,monatomic}}{c_{gas,diatomic}} = \sqrt{\frac{5/3}{7/5}} = \sqrt{\frac{25}{21}} = 1.091...$$

This gives the 9% difference, and would be a typical ratio for speeds of sound at room temperature in helium vs. deuterium, each with a molecular weight of 4. Sound travels faster in helium than deuterium because adiabatic compression heats helium more, since the helium molecules can store heat energy from compression only in translation, but not rotation. Thus helium molecules (monatomic molecules) travel faster in a sound wave and transmit sound faster. (Sound travels at about 70% of the mean molecular speed in gases; the figure is 75% in monatomic gases and 68% in diatomic gases).

Note that in this example we have assumed that temperature is low enough that heat capacities are not influenced by molecular vibration (see heat capacity). However, vibrational modes simply cause gammas which decrease toward 1, since vibration modes in a polyatomic gas gives the gas additional ways to store heat which do not affect temperature, and thus do not affect molecular velocity and sound velocity. Thus, the effect of higher temperatures and vibrational heat capacity acts to increase the difference between the speed of sound in monatomic vs. polyatomic molecules, with the speed remaining greater in monatomics.

Practical Application to Air

By far the most important factor influencing the speed of sound in air is temperature. The speed is proportional to the square root of the absolute temperature, giving an increase of about 0.6 m/s per degree Celsius. For this reason, the pitch of a musical wind instrument increases as its temperature increases.

The speed of sound is raised by humidity but decreased by carbon dioxide. The difference between 0% and 100% humidity is about 1.5 m/s at standard pressure and temperature, but the size of the humidity effect increases dramatically with temperature. The carbon dioxide content of air is not fixed, due to both carbon pollution and human breath (e.g., in the air blown through wind instruments).

The dependence on frequency and pressure are normally insignificant in practical applications. In dry air, the speed of sound increases by about 0.1 m/s as the frequency rises from 10 Hz to 100 Hz. For audible frequencies above 100 Hz it is relatively constant. Standard values of the speed of sound are quoted in the limit of low frequencies, where the wavelength is large compared to the mean free path.

Mach Number

Mach number, a useful quantity in aerodynamics, is the ratio of air speed to the local speed of sound. At altitude, for reasons explained, Mach number is a function of temperature. Aircraft flight instruments, however, operate using pressure differential to compute Mach number, not temperature. The assumption is that a particular pressure represents a particular altitude and, therefore, a standard temperature. Aircraft flight instruments need to operate this way because the stagnation pressure sensed by a Pitot tube is dependent on altitude as well as speed.

Experimental Methods

A range of different methods exist for the measurement of sound in air.

The earliest reasonably accurate estimate of the speed of sound in air was made by William Derham, and acknowledged by Isaac Newton. Derham had a telescope at the top of the tower of the Church of St Laurence in Upminster, England. On a calm day, a synchronized pocket watch would be given to an assistant who would fire a shotgun at a pre-determined time from a conspicuous point some miles away, across the countryside. This could be confirmed by telescope. He then measured the interval between seeing gunsmoke and arrival of the sound using a half-second pendulum. The distance from where the gun was fired was found by triangulation, and simple division (distance/time) provided velocity. Lastly, by making many observations, using a range of different distances, the inaccuracy of the half-second pendulum could be averaged out, giving his final estimate of the speed of sound. Modern stopwatches enable this method to be used today over distances as short as 200–400 meters, and not needing something as loud as a shotgun.

Single-shot Timing Methods

The simplest concept is the measurement made using two microphones and a fast recording device such as a digital storage scope. This method uses the following idea.

If a sound source and two microphones are arranged in a straight line, with the sound source at one end, then the following can be measured:

1. The distance between the microphones (x), called microphone basis.

2. The time of arrival between the signals (delay) reaching the different microphones (t).

Then $v = x/t$.

Other Methods

In these methods the time measurement has been replaced by a measurement of the inverse of time (frequency).

Kundt's tube is an example of an experiment which can be used to measure the speed of sound in a small volume. It has the advantage of being able to measure the speed of sound in any gas. This method uses a powder to make the nodes and antinodes visible to the human eye. This is an example of a compact experimental setup.

A tuning fork can be held near the mouth of a long pipe which is dipping into a barrel of water. In this system it is the case that the pipe can be brought to resonance if the length of the air column in the pipe is equal to $(1 + 2n)\lambda/4$ where n is an integer. As the antinodal point for the pipe at the open end is slightly outside the mouth of the pipe it is best to find two or more points of resonance and then measure half a wavelength between these.

Here it is the case that $v = f\lambda$.

High-precision Measurements in Air

The effect from impurities can be significant when making high-precision measurements. Chemical desiccants can be used to dry the air, but will in turn contaminate the sample. The air can be dried cryogenically, but this has the effect of removing the carbon dioxide as well; therefore many high-precision measurements are performed with air free of carbon dioxide rather than with natural air. A 2002 review found that a 1963 measurement by Smith and Harlow using a cylindrical resonator gave "the most probable value of the standard speed of sound to date." The experiment was done with air from which the carbon dioxide had been removed, but the result was then corrected for this effect so as to be applicable to real air. The experiments were done at 30 °C but corrected for temperature in order to report them at 0 °C. The result was 331.45 ± 0.01 m/s for dry air at STP, for frequencies from 93 Hz to 1,500 Hz.

Speed of Sound in Solids

Three-dimensional Solids

In a solid, there is a non-zero stiffness both for volumetric deformations and shear deformations.

Hence, it is possible to generate sound waves with different velocities dependent on the deformation mode. Sound waves generating volumetric deformations (compression) and shear deformations (shearing) are called pressure waves (longitudinal waves) and shear waves (transverse waves), respectively. In earthquakes, the corresponding seismic waves are called P-waves (primary waves) and S-waves (secondary waves), respectively. The sound velocities of these two types of waves propagating in a homogeneous 3-dimensional solid are respectively given by

$$c_{\text{solid,p}} = \sqrt{\frac{K + \frac{4}{3}G}{\rho}} = \sqrt{\frac{E(1-v)}{\rho(1+v)(1-2v)}},$$

$$c_{\text{solid,s}} = \sqrt{\frac{G}{\rho}},$$

where

- K is the bulk modulus of the elastic materials;
- G is the shear modulus of the elastic materials;
- E is the Young's modulus;
- ρ is the density;
- v is Poisson's ratio.

The last quantity is not an independent one, as E = 3K(1 − 2v). Note that the speed of pressure waves depends both on the pressure and shear resistance properties of the material, while the speed of shear waves depends on the shear properties only.

Typically, pressure waves travel faster in materials than do shear waves, and in earthquakes this is the reason that the onset of an earthquake is often preceded by a quick upward-downward shock, before arrival of waves that produce a side-to-side motion. For example, for a typical steel alloy, K = 170 GPa, G = 80 GPa and ρ = 7,700 kg/m³, yielding a compressional speed $c_{\text{solid,p}}$ of 6,000 m/s. This is in reasonable agreement with $c_{\text{solid,p}}$ measured experimentally at 5,930 m/s for a (possibly different) type of steel. The shear speed $c_{\text{solid,s}}$ is estimated at 3,200 m/s using the same numbers.

One-dimensional Solids

The speed of sound for pressure waves in stiff materials such as metals is sometimes given for "long rods" of the material in question, in which the speed is easier to measure. In rods where their diameter is shorter than a wavelength, the speed of pure pressure waves may be simplified and is given by:

$$c_{\text{solid}} = \sqrt{\frac{E}{\rho}},$$

where E is the Young's modulus. This is similar to the expression for shear waves, save that Young's modulus replaces the shear modulus. This speed of sound for pressure waves in long rods will always be slightly less than the same speed in homogeneous 3-dimensional solids, and the ratio of the speeds in the two different types of objects depends on Poisson's ratio for the material.

Speed of Sound in Liquids

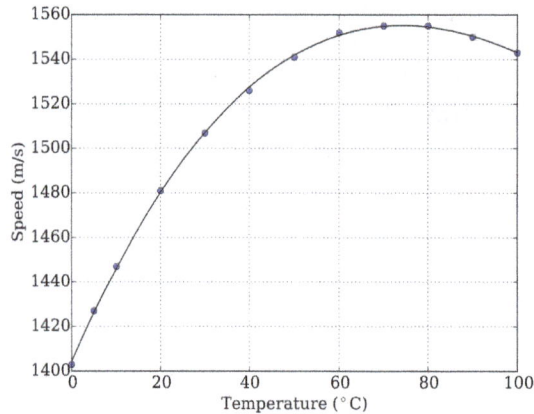

Speed of sound in water vs temperature.

In a fluid the only non-zero stiffness is to volumetric deformation (a fluid does not sustain shear forces).

Hence the speed of sound in a fluid is given by

$$c_{\text{fluid}} = \sqrt{\frac{K}{\rho}},$$

where K is the bulk modulus of the fluid.

Water

In fresh water, sound travels at about 1481 m/s at 20 °C. Applications of underwater sound can be found in sonar, acoustic communication and acoustical oceanography.

Seawater

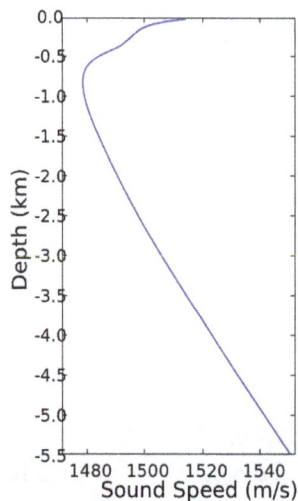

Speed of sound as a function of depth at a position north of Hawaii in the Pacific Ocean derived from the 2005 World Ocean Atlas. The SOFAR channel is centred on the minimum in the speed of sound at ca. 750-m depth.

In salt water that is free of air bubbles or suspended sediment, sound travels at about 1500 m/s (1500.235 m/s at 1000 kilopascals, 10 °C and 3% salinity by one method). The speed of sound in seawater depends on pressure (hence depth), temperature (a change of 1 °C ~ 4 m/s), and salinity (a change of 1‰ ~ 1 m/s), and empirical equations have been derived to accurately calculate the speed of sound from these variables. Other factors affecting the speed of sound are minor. Since temperature decreases with depth while pressure and generally salinity increase, the profile of the speed of sound with depth generally shows a characteristic curve which decreases to a minimum at a depth of several hundred meters, then increases again with increasing depth (right). For more information see Dushaw et al.

A simple empirical equation for the speed of sound in sea water with reasonable accuracy for the world's oceans is due to Mackenzie:

$$c(T,S,z) = a_1 + a_2T + a_3T^2 + a_4T^3 + a_5(S-35) + a_6z + a_7z^2 + a_8T(S-35) + a_9Tz^3,$$

where

- T is the temperature in degrees Celsius;

- S is the salinity in parts per thousand;

- z is the depth in meters.

The constants $a_1, a_2, ..., a_9$ are

$$
\begin{aligned}
a_1 &= 1,448.96, & a_2 &= 4.591, & a_3 &= -5.304 \times 10^{-2}, \\
a_4 &= 2.374 \times 10^{-4}, & a_5 &= 1.340, & a_6 &= 1.630 \times 10^{-2}, \\
a_7 &= 1.675 \times 10^{-7}, & a_8 &= -1.025 \times 10^{-2}, & a_9 &= -7.139 \times 10^{-13},
\end{aligned}
$$

with check value 1550.744 m/s for $T = 25$ °C, $S = 35$ parts per thousand, $z = 1,000$ m. This equation has a standard error of 0.070 m/s for salinity between 25 and 40 ppt. Speed of Sound in Sea-Water for an online calculator.

Other equations for the speed of sound in sea water are accurate over a wide range of conditions, but are far more complicated, e.g., that by V. A. Del Grosso and the Chen-Millero-Li Equation.

Speed of Sound in Plasma

The speed of sound in a plasma for the common case that the electrons are hotter than the ions (but not too much hotter) is given by the formula

$$c_s = (\gamma Z k T_e / m_i)^{1/2} = 9.79 \times 10^3 (\gamma Z T_e / \mu)^{1/2} \text{ m/s,}$$

where

- m_i is the ion mass;

- μ is the ratio of ion mass to proton mass $\mu = m_i/m_p$;

- T_e is the electron temperature;

- Z is the charge state;

- k is Boltzmann constant;

- γ is the adiabatic index.

In contrast to a gas, the pressure and the density are provided by separate species, the pressure by the electrons and the density by the ions. The two are coupled through a fluctuating electric field.

Gradients

When sound spreads out evenly in all directions in three dimensions, the intensity drops in proportion to the inverse square of the distance. However, in the ocean there is a layer called the 'deep sound channel' or SOFAR channel which can confine sound waves at a particular depth.

In the SOFAR channel, the speed of sound is lower than that in the layers above and below. Just as light waves will refract towards a region of higher index, sound waves will refract towards a region where their speed is reduced. The result is that sound gets confined in the layer, much the way light can be confined in a sheet of glass or optical fiber. Thus, the sound is confined in essentially two dimensions. In two dimensions the intensity drops in proportion to only the inverse of the distance. This allows waves to travel much further before being undetectably faint.

A similar effect occurs in the atmosphere. Project Mogul successfully used this effect to detect a nuclear explosion at a considerable distance.

Longitudinal Wave

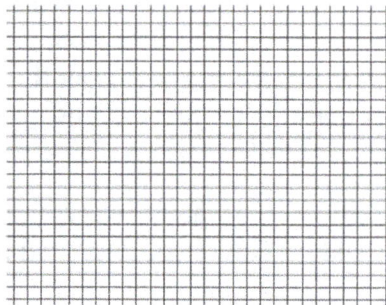

Plane pressure pulse wave

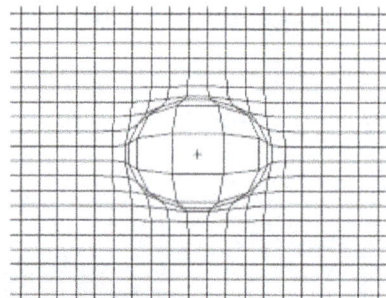

Representation of the propagation of an omnidirectional pulse wave on a 2d grid (empirical shape)

Longitudinal waves are waves in which the displacement of the medium is in the same direction as, or the opposite direction to, the direction of propagation of the wave. Mechanical longitudinal waves are also called *compressional* or *compression waves,* because they produce compression and rarefaction when traveling through a medium, and *pressure waves,* because they produce increases and decreases in pressure. The other main type of wave is the transverse/transversal wave, in which the displacements of the medium are at right angles to the direction of propagation. Some transverse waves are mechanical, meaning that the wave needs a medium to travel through. Transverse mechanical waves are also called "shear waves".

By acronym, "longitudinal waves" and "transverse waves" were occasionally abbreviated by some authors as "L-waves" and "T-waves" respectively for their own convenience. While these two acronyms have specific meanings in seismology (L-wave for Love wave or long wave) and electrocardiography (see T wave), some authors chose to use "*l-waves*" (lowercase 'L') and "*t-waves*" instead, although they are not commonly found in physics writings except for some popular science books.

Examples

Longitudinal waves include sound waves (vibrations in pressure, particle of displacement, and particle velocity propagated in an elastic medium) and seismic P-waves (created by earthquakes and explosions). In longitudinal waves, the displacement of the medium is parallel to the propagation of the wave. A wave along the length of a stretched Slinky toy, where the distance between coils increases and decreases, is a good visualization. Sound waves in air are longitudinal, pressure waves.

Sound Waves

In the case of longitudinal harmonic sound waves, the frequency and wavelength can be described by the formula

$$y(x,t) = y_0 \cos\left(\omega\left(t - \frac{x}{c}\right)\right)$$

where:

- y is the displacement of the point on the traveling sound wave;
- x is the distance the point has traveled from the wave's source;
- t is the time elapsed;
- y_0 is the amplitude of the oscillations,
- c is the speed of the wave; and
- ω is the angular frequency of the wave.

The quantity x/c is the time that the wave takes to travel the distance x.

The ordinary frequency (f) of the wave is given by

$$f = \frac{\omega}{2\pi}.$$

The wavelength can be calculated as the relation between a wave's speed and ordinary frequency.

$$\lambda = \frac{c}{f}.$$

For sound waves, the amplitude of the wave is the difference between the pressure of the undisturbed air and the maximum pressure caused by the wave.

Sound's propagation speed depends on the type, temperature, and composition of the medium through which it propagates.

Pressure Waves

In an elastic medium with rigidity, a harmonic pressure wave oscillation has the form,

$$y(x,t) = y_0 \cos(kx - \omega t + \varphi)$$

where:

- y_0 is the amplitude of displacement,
- k is the wavenumber,
- x is the distance along the axis of propagation,
- ω is the angular frequency,
- t is the time, and
- φ is the phase difference.

The restoring force, which acts to return the medium to its original position, is provided by the medium's bulk modulus.

Electromagnetic

Maxwell's equations lead to the prediction of electromagnetic waves in a vacuum, which are transverse (in that the electric fields and magnetic fields vary perpendicularly to the direction of propagation). However, waves can exist in plasmas or confined spaces, called plasma waves, which can be longitudinal, transverse, or a mixture of both. Plasma waves can also occur in force-free magnetic fields.

In the early development of electromagnetism, there were some like Alexandru Proca (1897-1955) known for developing relativistic quantum field equations bearing his name (Proca's equations) for the massive, vector spin-1 mesons. In recent decades some extended electromagnetic theorists, such as Jean-Pierre Vigier and Bo Lehnert of the Swedish Royal Society, have used the Proca equation in an attempt to demonstrate photon mass as a longitudinal electromagnetic component of Maxwell's equations, suggesting that longitudinal electromagnetic waves could exist in a Dirac polarized vacuum.

After Heaviside's attempts to generalize Maxwell's equations, Heaviside came to the conclusion that electromagnetic waves were not to be found as longitudinal waves in *"free space"* or homogeneous media. But Maxwell's equations do lead to the appearance of longitudinal waves under some circumstances, for example, in plasma waves or guided waves. Basically distinct from the "free-space" waves, such as those studied by Hertz in his UHF experiments, are Zenneck waves. The longitudinal modes of a resonant cavity are the particular standing wave patterns formed by waves confined in a cavity. The longitudinal modes correspond to those wavelengths of the wave which are reinforced by constructive interference after many reflections from the cavity's reflecting surfaces. Recently, Haifeng Wang et al. proposed a method that can generate a longitudinal electromagnetic (light) wave in free space, and this wave can propagate without divergence for a few wavelengths.

Timbre

In music, timbre (also known as tone color or tone quality from psychoacoustics) is the perceived sound quality of a musical note, sound, or tone that distinguishes different types of sound production, such as choir voices and musical instruments, such as string instruments, wind instruments, and percussion instruments, and which enables listeners to hear even different instruments from the same category as different (e.g. a viola and a violin).

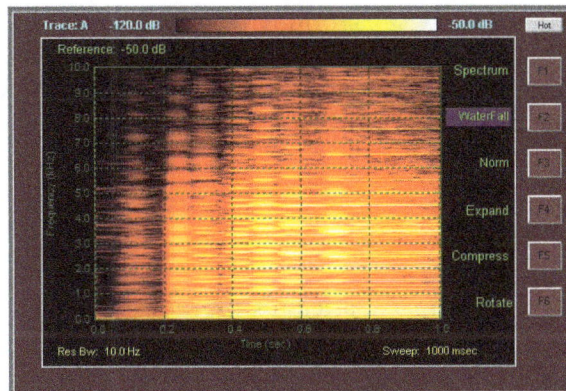

Spectrogram of the first second of an E9 suspended chord played
on a Fender Stratocaster guitar with noiseless pickups. Below is the E9 suspended chord audio:

The physical characteristics of sound that determine the perception of timbre include spectrum and envelope. Singers and instrumental musicians can change the timbre of the music they are singing/playing by using different singing or playing techniques. For example, a violinist can use different bowing styles or play on different parts of the string to obtain different timbres (e.g., playing sul tasto produces a light, airy timbre, whereas playing sul ponticello produces a harsh, even an aggressive tone). On electric guitar and electric piano, performers can change the timbre using effects units and graphic equalizers.

In simple terms, timbre is what makes a particular musical sound have a different sound from another, even when they have the same pitch and loudness. For instance, it is the difference in sound between a guitar and a piano playing the same note at the same volume. Both instruments can sound equally tuned in relation to each other as they play the same note, and while playing at the same amplitude level each instrument will still sound distinctively with its own unique tone color. Experienced musicians are able to distinguish between different instruments of the same type based on their varied timbres, even if those instruments are playing notes at the same pitch and loudness.

Synonyms

Tone quality and *tone color* are synonyms for *timbre*, as well as the "*texture* attributed to a single instrument". However, the word texture can also refer to the type of music, such as multiple, interweaving melody lines versus a singable melody accompanied by subordinate chords. Hermann von Helmholtz used the German *Klangfarbe* (*tone color*), and John Tyndall proposed an English translation, *clangtint*, but both terms were disapproved of by Alexander Ellis, who also discredits *register* and *color* for their pre-existing English meanings (Erickson 1975, 7). The sound of a musical instrument may be described with words such as *bright*, *dark*, *warm*, *harsh*, and other terms. There are also colors of noise, such as pink and white. In visual representations of sound, timbre corresponds to the shape of the image (Abbado 1988, 3), while loudness corresponds to brightness; pitch corresponds to the y-shift of the spectogram.

ASA Definition

The Acoustical Society of America (ASA) Acoustical Terminology definition 12.09 of timbre describes it as "that attribute of auditory sensation which enables a listener to judge that two non-identical sounds, similarly presented and having the same loudness and pitch, are dissimilar", adding, "Timbre depends primarily upon the frequency spectrum, although it also depends upon the sound pressure and the temporal characteristics of the sound" (Acoustical Society of America Standards Secretariat 1994).

Attributes

Timbre has been called, "...the psychoacoustician's multidimensional waste-basket category for everything that cannot be labeled pitch or loudness." (McAdams and Bregman 1979, 34; *cf.* Dixon Ward 1965, 55 and Tobias 1970, 409).

Many commentators have attempted to decompose timbre into component attributes. For example, J. F. Schouten (1968, 42) describes the, "elusive attributes of timbre", as "determined by at least five major acoustic parameters", which Robert Erickson (1975, 5) finds, "scaled to the concerns of much contemporary music":

1. The range between tonal and noiselike character

2. The spectral envelope

3. The time envelope in terms of rise, duration, and decay (ADSR—attack, decay, sustain, release)

4. The changes both of spectral envelope (formant-glide) and fundamental frequency (micro-intonation)

5. The prefix, or onset of a sound, quite dissimilar to the ensuing lasting vibration

Erickson (1975, 6) gives a table of subjective experiences and related physical phenomena based on Schouten's five attributes:

Subjective	Objective
Tonal character, usually pitched	Periodic sound
Noisy, with or without some tonal character, including rustle noise	Noise, including random pulses characterized by the rustle time (the mean interval between pulses)
Coloration	Spectral envelope
Beginning/ending	Physical rise and decay time
Coloration glide or formant glide	Change of spectral envelope
Microintonation	Small change (one up and down) in frequency
Vibrato	Frequency modulation
Tremolo	Amplitude modulation
Attack	Prefix
Final sound	Suffix

Harmonics

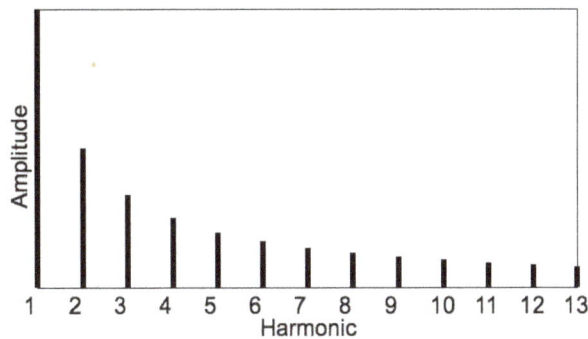

Harmonic spectra.

The richness of a sound or note a musical instrument produces is sometimes described in terms of a sum of a number of distinct frequencies. The lowest frequency is called the *fundamental frequency*, and the pitch it produces is used to name the note, but the fundamental frequency is not always the dominant frequency. The dominant frequency is the frequency that is most heard, and it is always a multiple of the fundamental frequency. For example, the dominant frequency for the transverse flute is double the fundamental frequency. Other significant frequencies are called overtones of the fundamental frequency, which may include harmonics and partials. Harmonics are whole number multiples of the fundamental frequency, such as ×2, ×3, ×4, etc. Partials are other overtones. There are also sometimes subharmonics at whole number *divisions* of the fundamental frequency. Most instruments produce harmonic sounds, but many instruments produce partials and inharmonic tones, such as cymbals and other indefinite-pitched instruments.

When the tuning note in an orchestra or concert band is played, the sound is a combination of 440 Hz, 880 Hz, 1320 Hz, 1760 Hz and so on. Each instrument in the orchestra or concert band produces a different combination of these frequencies, as well as harmonics and overtones. The sound waves of the different frequencies overlap and combine, and the balance of these amplitudes is a major factor in the characteristic sound of each instrument.

William Sethares wrote that just intonation and the western equal tempered scale are related to the harmonic spectra/timbre of many western instruments in an analogous way that the inharmonic timbre of the Thai renat (a xylophone-like instrument) is related to the seven-tone near-equal tempered pelog scale in which they are tuned. Similarly, the inharmonic spectra of Balinese metallophones combined with harmonic instruments such as the stringed rebab or the voice, are related to the five-note near-equal tempered slendro scale commonly found in Indonesian gamelan music (Sethares 1998, 6, 211, 318).

Envelope

The timbre of a sound is also greatly affected by the following aspects of its *envelope:* attack time and characteristics, decay, sustain, release (ADSR envelope) and transients. Thus these are all common controls on synthesizers. For instance, if one takes away the attack from the sound of a piano or trumpet, it becomes more difficult to identify the sound correctly, since the sound of the hammer hitting the strings or the first blast of the player's lips are highly characteristic of those instruments. The envelope is the overall amplitude structure of a sound, so called because the sound just "fits" inside its envelope: what this means should be clear from a time-domain display of almost any interesting sound, zoomed out enough that the entire waveform is visible.

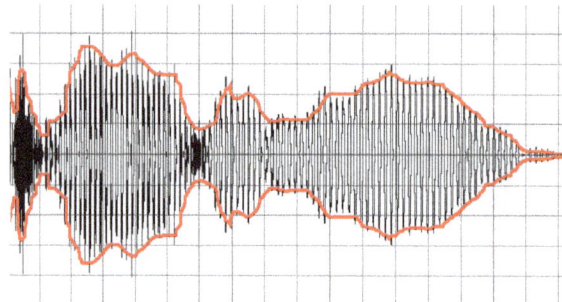

A signal and its envelope marked with red

In Music History

Instrumental timbre played an increasing role in the practice of orchestration during the eighteenth and nineteenth centuries. Berlioz (Macdonald 1969, 51) and Wagner (Latham) made significant contributions to its development during the nineteenth century. Debussy, who composed during the last decades of the nineteenth and the first decades of the twentieth centuries, has been credited with elevating further the role of timbre: "To a marked degree the music of Debussy elevates timbre to an unprecedented structural status; already in *Prélude à l'après-midi d'un faune* the *color* of flute and harp functions referentially" (Samson 1977, p. 195).

In rock music from the late 1960s to the 2000s, the timbre of specific sounds is important to a song. For example, in heavy metal music, the sonic impact of the heavily-amplified, heavily-distorted power chord played on electric guitar through very loud guitar amplifiers and rows of speaker cabinets is an essential part of the style's musical identity. You could not take the heavily-amplified electric guitar part and substitute it with the same notes played on a grand piano or pipe organ.

Psychoacoustic Evidence

Often, listeners can identify an instrument, even at different pitches and loudness, in different environments, and with different players. In the case of the clarinet, acoustic analysis shows waveforms irregular enough to suggest three instruments rather than one. David Luce (1963, 16) suggests that this implies that, "Certain strong regularities in the acoustic waveform of the above instruments must exist which are invariant with respect to the above variables." However, Robert Erickson argues that there are few regularities and they do not explain our "...powers of recognition and identification." He suggests borrowing the concept of subjective constancy from studies of vision and visual perception (Erickson 1975, 11).

Psychoacoustic experiments from the 1960s onwards tried to elucidate the nature of timbre. One method involves playing pairs of sounds to listeners, then using a multidimensional scaling algorithm to aggregate their dissimilarity judgments into a timbre space. The most consistent outcomes from such experiments are that brightness or spectral energy distribution (Grey 1977), and the *bite*, or rate and synchronicity (Wessel 1979) and rise time (Lakatos 2000), of the attack are important factors.

Tristimulus Timbre Model

The concept of tristimulus originates in the world of color, describing the way three primary colors can be mixed together to create a given color. By analogy, the musical tristimulus measures the mixture of harmonics in a given sound, grouped into three sections. The first tristimulus measures the relative weight of the first harmonic; the second tristimulus measures the relative weight of the second, third, and fourth harmonics taken together; and the third tristimulus measures the relative weight of all the remaining harmonics (Peeters 2003; Pollard and Jansson 1982,):

$$T_1 = \frac{a_1}{\sum_{h=1}^{H} a_h} \qquad T_2 = \frac{a_2 + a_3 + a_4}{\sum_{h=1}^{H} a_h} \qquad T_3 = \frac{\sum_{h=5}^{H} a_h}{\sum_{h=1}^{H} a_h}$$

Loudness

Loudness is the characteristic of a sound that is primarily a psycho-physiological correlate of physical strength (amplitude). More formally, it is defined as, "That attribute of auditory sensation in terms of which sounds can be ordered on a scale extending from quiet to loud." The relation of physical attributes of sound to perceived loudness consists of physical, physiological and psychological components.

In different industries, loudness may have different meanings, and different standards exist, each purporting to define the measurement. Some definitions such as LKFS refer to relative loudness of different segments of electronically reproduced sounds such as for broadcasting and cinema. Others, such as ISO 532A (Stevens loudness, measured in sones), ISO 532B (Zwicker loudness), DIN

45631 and ASA/ANSI S3.4, have a more general scope and are often used to characterize loudness of environmental noise.

It is sometimes stated that loudness is a subjective measure, often confused with physical measures of sound strength such as sound pressure, sound pressure level (in decibels), sound intensity or sound power. It is often possible to separate the truly subjective components such as social considerations from the physical and physiological.

Filters such as A-weighting attempt to adjust sound measurements to correspond to loudness as perceived by the typical human, however this approach is only truly valid for loudness of single tones. A-weighting follows human sensitivity to sound and describes relative perceived loudness for at quiet to moderate speech levels, around 40 phons. However, physiological loudness perception is a much more complex process than can be captured with a single correction curve. Not only do equal-loudness contours vary with intensity, but perceived loudness of a complex sound depends on whether its spectral components are closely or widely spaced in frequency. When generating neural impulses in response to sounds of one frequency, the ear is less sensitive to nearby frequencies, which are said to be in the same critical band. Sounds containing spectral components in many critical bands are perceived as louder even if the total sound pressure remains constant.

Explanation

The perception of loudness is related to sound pressure level (SPL), frequency content and duration of a sound. The human auditory system averages the effects of SPL over a 600–1000 ms interval. A sound of constant SPL will be perceived to increase in loudness as samples of duration 20, 50, 100, 200 ms are heard, up to a duration of about 1 second at which point the perception of loudness will stabilize. For sounds of duration greater than 1 second, the moment-by-moment perception of loudness will be related to the average loudness during the preceding 600–1000 ms.

For sounds having a duration longer than 1 second, the relationship between SPL and loudness of a single tone can be approximated by Stevens' power law in which SPL has an exponent of 0.6. More precise measurements indicate that loudness increases with a higher exponent at low and high levels and with a lower exponent at moderate levels.

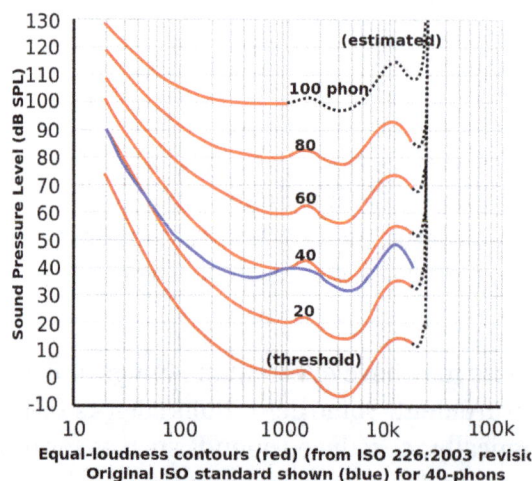

Equal-loudness contours (red) (from ISO 226:2003 revisic
Original ISO standard shown (blue) for 40-phons

The horizontal axis shows frequency in Hz

The sensitivity of the human ear changes as a function of frequency, as shown in the equal-loudness graph. Each line on this graph shows the SPL required for frequencies to be perceived as equally loud, and different curves pertain to different sound pressure levels. It also shows that humans with normal hearing are most sensitive to sounds around 2–4 kHz, with sensitivity declining to either side of this region. A complete model of the perception of loudness will include the integration of SPL by frequency.

Historically, loudness was measured using an "ear-balance" audiometer in which the amplitude of a sine wave was adjusted by the user to equal the perceived loudness of the sound being evaluated. Contemporary standards for measurement of loudness are based on summation of energy in critical bands as described in IEC 532, DIN 45631 and ASA/ANSI S3.4. A distinction is made between stationary loudness (sounds that remain sensibly constant) and non-stationary (sound sources that move in space or change amplitude over time.)

Hearing Loss

When sensorineural hearing loss (damage to the cochlea or in the brain) is present, the perception of loudness is altered. Sounds at low levels (often perceived by those without hearing loss as relatively quiet) are no longer audible to the hearing impaired, but sounds at high levels often are perceived as having the same loudness as they would for an unimpaired listener. This phenomenon can be explained by two theories: loudness grows more rapidly for these listeners than normal listeners with changes in level. This theory is called "loudness recruitment" and has been accepted as the classical explanation. More recently, it has been proposed that some listeners with sensorineural hearing loss may in fact exhibit a normal rate of loudness growth, but instead have an elevated loudness at their threshold. That is, the softest sound that is audible to these listeners is louder than the softest sound audible to normal listeners. This theory is called "softness imperception", a term coined by Mary Florentine.

Compensation

The "loudness" control on some consumer stereos alters the frequency response curve to correspond roughly with the equal loudness characteristic of the ear. Loudness compensation is intended to make the recorded music sound more natural when played at a lower levels by boosting low frequencies, to which the ear is less sensitive at lower sound pressure levels.

Normalization

Loudness normalization is a specific type of audio normalization that equalizes perceived level such that, for instance, commercials do not sound louder than television programs. Loudness normalization schemes exist for a number of audio applications.

Broadcast

- Commercial Advertisement Loudness Mitigation Act
- European Broadcasting Union R128

Movie and Home Theaters

- Dialnorm

Music Playback

- Sound Check in iTunes
- ReplayGain

Measurement

Historically Sone (loudness N) and Phon (loudness level L) units have been used to measure loudness.

Relative loudness monitoring in production is measured in accordance with ITU-R BS.1770 in units of LKFS.

Work began on ITU-R BS.1770 in 2001 after 0 dBFS+ level distortion in converters and lossy codecs had become evident; and the original Leq(RLB) loudness metric was proposed by Gilbert Soloudre in 2003.

Based on data from subjective listening tests, Leq(RLB) was compared against numerous other algorithms where it did remarkably well. After modification of the frequency weighting, the measurement was made multi-channel (monaural to 5.1 surround sound). CBC, Dolby and TC Electronics and numerous broadcasters contributed to the listening tests.

To make the loudness metric cross-genre friendly, a relative measurement gate was added. This work was carried out in 2008 by the EBU. The improvements were brought back into BS.1770-2. ITU subsequently updated the true-peak metric (BS.1770-3) and added provision for more audio channels, for instance 22.2 surround sound (BS.1770-4).

Soundscape

The soundscape is the component of the acoustic environment that can be perceived by humans. There is a varied history of the use of soundscape depending on discipline - ranging from urban design to wildlife ecology. An important distinction is to separate soundscape from the broader term acoustic environment. The acoustic environment is the combination of all the acoustic resources within a given area - natural sounds and human-caused sounds – as modified by the environment. The International Organization for Standardization (ISO) standardized these definitions in 2014.

Historical Context

The term soundscape was first noted by Michael Southworth in a 1969 article titled "The Sonic Environment of Cities," published by Environment and Behavior, and fleshed out in more detail eight years later by Canadian composer and naturalist, R. Murray Schafer in his seminal work, "Tuning of the World." According to this author there are three main elements of the soundscape:

A soundscape is a sound or combination of sounds that forms or arises from an immersive environment. The study of soundscape is the subject of acoustic ecology or soundscape ecology. The idea of soundscape refers to both the natural acoustic environment, consisting of natural sounds, including animal vocalizations, the collective habitat expression of which is now referred to as the biophony, and, for instance, the sounds of weather and other natural elements, now referred to as the geophony; and environmental sounds created by humans, the anthropophony through a sub-set called controlled sound, such as musical composition, sound design, and language, work, and sounds of mechanical origin resulting from use of industrial technology. Crucially, the term soundscape also includes the listener's *perception* of sounds heard as an environment: "how that environment is understood by those living within it" and therefore mediates their relations. The disruption of these acoustic environments results in noise pollution.

The term "soundscape" can also refer to an audio recording or performance of sounds that create the sensation of experiencing a particular acoustic environment, or compositions created using the found sounds of an acoustic environment, either exclusively or in conjunction with musical performances.

Pauline Oliveros, composer of post-World War II electronic art music, defined the term "soundscape" as "All of the waveforms faithfully transmitted to our audio cortex by the ear and its mechanisms".

- Keynote sounds

    This is a musical term that identifies the key of a piece, not always audible ... the key might stray from the original, but it will return. The keynote sounds may not always be heard consciously, but they "outline the character of the people living there" (Schafer). They are created by nature (geography and climate): wind, water, forests, plains, birds, insects, animals. In many urban areas, traffic has become the keynote sound.

- Sound signals

    These are foreground sounds, which are listened to consciously; examples would be warning devices, bells, whistles, horns, sirens, etc.

- Soundmark

    This is derived from the term *landmark*. A soundmark is a sound which is unique to an area. In his 1993 book, *The Soundscape: Our Sonic Environment and the Tuning of the World*, Schafer wrote, "Once a Soundmark has been identified, it deserves to be protected, for soundmarks make the acoustic life of a community unique."

The elements have been further defined as to essential sources:

Bernie Krause, naturalist and soundscape ecologist, redefined the sources of sound in terms of their three main components: geophony, biophony, and anthropophony.

- Geophony

    Consisting of the prefix, geo (gr. earth), and phon (gr. sound), this refers to the soundscape sources that are generated by non-biological natural sources such as wind in the trees, water in a stream or waves at the ocean, and earth movement, the first sounds heard on earth by any sound-sentient organism.

- Biophony

  Consisting of the prefix, bio (gr. life) and the suffix for sound, this term refers to all of the non-human, non-domestic biological soundscape sources of sound.

- Anthropophony

  Consisting of the prefix, anthro (gr. human), this term refers to all of the sound signatures generated by humans.

In Music

In music, soundscape compositions are often a form of electronic music, or electroacoustic music. Composers who use soundscapes include real-time granular synthesis pioneer Barry Truax, Hildegard Westerkamp, and Luc Ferrari, whose *Presque rien, numéro 1* (1970) is an early soundscape composition. Soundscape composer Petri Kuljuntausta has created soundscape compositions from the sounds of sky dome and Aurora Borealis and deep sea underwater recordings, and a work entitled "Charm of Sound" to be performed at the extreme environment of Saturn's moon Titan. The work landed on the ground of Titan in 2005 after traveling inside the spacecraft *Huygens* over seven years and four billion kilometres through space.

Irv Teibel's *Environments series* (1969–79) consisted of 30-minute, uninterrupted environmental soundscapes and synthesized or processed versions of natural sound.

Music soundscapes can also be generated by automated software methods, such as the experimental TAPESTREA application, a framework for sound design and soundscape composition, and others.

The soundscape is often the subject of mimicry in Timbre-centered music such as Tuvan throat singing. The process of Timbral Listening is used to interpret the timbre of the soundscape. This timbre is mimicked and reproduced using the voice or rich harmonic producing instruments.

In United States National Parks

The National Park Service Natural Sounds and Night Skies Division actively protects the soundscapes and acoustic environments in national parks across the country. It is important to distinguish and define certain key terms as used by the National Park Service. *Acoustic resources* are physical sound sources, including both natural sounds (wind, water, wildlife, vegetation) and cultural and historic sounds (battle reenactments, tribal ceremonies, quiet reverence). The acoustic environment is the combination of all the acoustic resources within a given area - natural sounds and human-caused sounds – as modified by the environment. The acoustic environment includes sound vibrations made by geological processes, biological activity, and even sounds that are inaudible to most humans, such as bat echolocation calls. *Soundscape* is the component of the acoustic environment that can be perceived and comprehended by the humans. The character and quality of the soundscape influence human perceptions of an area, providing a sense of place that differentiates it from other regions. *Noise* refers to sound which is unwanted, either because of its effects on humans and wildlife, or its interference with the perception or detection of other sounds. *Cultural soundscapes* include opportunities for appropriate transmission of cultural and historic sounds that are fundamental components of the purposes and values for which the parks were established.

Soundscapes and the Environment

There are two distinct soundscapes, either hi-fi or lo-fi, created by the environment. A hi-fi system possesses a positive signal-to-noise ratio. These settings make it possible for discrete sounds to be heard clearly since there is no background noise to obstruct even the smallest disturbance. A rural landscape offers more hi-fi frequencies than a city because the natural landscape creates an opportunity to hear incidences from nearby and afar. In a lo-fi soundscape, signals are obscured by too many sounds, and perspective is lost within the broad- band of noises. In lo-fi soundscapes everything is very close and compact. A person can only listen to immediate encounters; in most cases even ordinary sounds have to be exuberantly amplified in order to be heard.

All sounds are unique in nature. They occur at one time in one place and can't be replicated. In fact, it is physically impossible for nature to reproduce any phoneme twice in exactly the same manner.

In Health Care

Soundscapes from a computerized acoustic device with a camera may also offer synthetic vision to the blind, utilizing human echolocation, as is the goal of the seeing with sound project.

Soundscapes and Noise Pollution

Papers on noise pollution are increasingly taking a holistic, soundscape approach to noise control. Whereas acoustics tends to rely on lab measurements and individual acoustic characteristics of cars and so on, soundscape takes a top-down approach. Drawing on John Cage's ideas of the whole world as composition, soundscape researchers investigate people's attitudes to soundscapes as a whole rather than individual aspects - and look at how the entire environment can be changed to be more pleasing to the ear.

It has been suggested that people's opportunity to access quiet, natural places in urban areas can be enhanced by improving the ecological quality of urban green spaces through targeted planning and design and that in turn has psychological benefits.

Soundscaping as a method to reducing noise polution incorporates natural elements rather than just man made elements.

Noise

Noise is unwanted sound judged to be unpleasant, loud or disruptive to hearing. From a physics standpoint, noise is indistinguishable from sound, as both are vibrations through a medium, such as air or water. The difference arises when the brain receives and perceives a sound.

NASA researchers at Glenn Research Center conducting tests on aircraft engine noise in 1967

In experimental sciences, noise can refer to any random fluctuations of data that hinders perception of an expected signal.

Acoustic noise is any sound in the acoustic domain, either deliberate (e.g., music or speech) or unintended. In contrast, noise in electronics may not be audible to the human ear and may require instruments for detection.

In audio engineering, noise can refer to the unwanted residual electronic noise signal that gives rise to acoustic noise heard as a hiss. This signal noise is commonly measured using A-weighting or ITU-R 468 weighting.

Measurement

Sound is measured based on the amplitude and frequency of a sound wave. Noise is most commonly discussed in terms of decibels (dB), the measure of loudness, or intensity of a sound; this measurement describes the amplitude of a sound wave. On the other hand, pitch describes the frequency of a sound and is measured in hertz (Hz).

Recording and Reproduction

In audio, recording, and broadcast systems, audio noise refers to the residual low-level sound (four major types: hiss, rumble, crackle, and hum) that is heard in quiet periods of program. This variation from the expected pure sound or silence can be caused by the audio recording equipment, the instrument, or ambient noise in the recording room.

In audio engineering it can refer either to the acoustic noise from loudspeakers or to the unwanted residual electronic noise signal that gives rise to acoustic noise heard as 'hiss'. This signal noise is commonly measured using A-weighting or ITU-R 468 weighting

Noise is often generated deliberately and used as a test signal for audio recording and reproduction equipment.

White Noise

White noise is energy randomly spread across a wide frequency band containing all notes from high to low. It is called "white" noise as it is analogous to "white" light which contains all the colors of the visible spectrum.

Environmental Noise

Environmental noise is the accumulation of all noise present in a specified environment. The principal sources of environmental noise are surface motor vehicles, aircraft, trains and industrial sources. These noise sources expose millions of people to noise pollution that creates not only annoyance, but also significant health consequences such as elevated incidence of hearing loss and cardiovascular disease. There are a variety of mitigation strategies and controls available to reduce sound levels including source intensity reduction, land use planning strategies, noise barriers and sound baffles, time of day use regimens, vehicle operational controls and architectural acoustics design measures.

Regulation of Noise

Certain geographic areas or specific occupations may be at a higher risk of being exposed to constantly high levels of noise; in order to prevent negative health outcomes, regulations may be set. Noise regulation includes statutes or guidelines relating to sound transmission established by national, state or provincial and municipal levels of government. Environmental noise is governed by laws and standards which set maximum recommended levels of noise for specific land uses, such as residential areas, areas of outstanding natural beauty, or schools. These standards usually specify measurement using a weighting filter, most often A-weighting.

United States

In 1972, the Noise Control Act was passed to promote a healthy living environment for all Americans, where noise does not pose a threat to human health. This policy's main objectives were: (1) establish coordination of research in the area of noise control, (2) establish federal standards on noise emission for commercial products, and (3) promote public awareness about noise emission and reduction.

The Quiet Communities Act of 1978 promotes noise control programs at the state and local level and developed a research program on noise control. Both laws authorized the Environmental Protection Agency to study the effects of noise and evaluate regulations regarding noise control.

Noise in the Workplace

In the US, the National Institute of Occupational Safety and Health (NIOSH) provides recommendation on noise exposure in the workplace. In 1972 (revised in 1998), NIOSH published a document outlining recommended standards relating to the occupational exposure to noise, with the purpose of reducing the risk of developing permanent hearing loss related to exposure at work. This publication set the recommended exposure limit (REL) of noise in an occupation setting to 85 dBA for 8 hours. However, in 1973 the Occupational Safety and Health Administration (OSHA) maintained the requirement of an 8-hour average of 90 dBA. The following year, OSHA required employers to provide a hearing conservation program to workers exposed to 85 dBA average 8-hour workdays.

Europe

The European Environment Agency regulates noise control and surveillance within the European Union. The Environmental Noise Directive was set to determine levels of noise exposure, increase public access to information regarding environmental noise, and reduce environmental noise. Additionally, in the European Union, underwater noise is a pollutant according to the Marine Strategy Framework Directive (MSFD). The MSFD requires EU Member States to achieve or maintain Good Environmental Status, meaning that the "introduction of energy, including underwater noise, is at levels that do not adversely affect the marine environment".

Health Effects from Noise

Exposure to noise is associated with several negative health outcomes. Depending on duration and level of exposure, noise may cause or increase the likelihood of hearing loss, high blood pressure, ischemic heart disease, sleep disturbances, injuries, and even decreased school performance.

Earplugs can be used to protect the user's ears from loud noises.

Noise exposure has increasingly been identified as a public health issue, especially in an occupational setting, as demonstrated with the creation of NIOSH's Noise and Hearing Loss Prevention program. Noise has also proven to be an occupational hazard, as it is the most common work-related pollutant. Noise-induced hearing loss, when associate with exposures from the workplace is also called occupational hearing loss.

Prevention

While noise-induced hearing loss is permanent, it is also very preventable. Particularly in the workplace, regulations may exist depicting maximum allowed levels of noise. This can be especially important for professionals working in settings with consistent exposure to loud sounds, such as musicians, music teachers and sound engineers. Examples of measures taken to prevent noise-induced hearing loss in the workplace include engineering noise control, the Buy-Quiet initiative, creation of the Safe-in-Sound Award, and noise surveillance.

Literary Views

Roland Barthes distinguishes between physiological noise, which is merely heard, and psychological noise, which is actively listened to. Physiological noise is felt subconsciously as the vibrations of the noise (sound) waves physically interact with the body while psychological noise is perceived as our conscious awareness shifts its attention to that noise.

Luigi Russolo, one of the first composers of noise music, wrote the essay *The Art of Noises*. He argues that any kind of noise could be used as music, as audiences become more familiar with noises caused by technological advancements; noise has become so prominent that pure sound no longer exists.

Henry Cowell claims that technological advancements have reduced unwanted noises from machines, but have not managed so far to completely eliminate them.

References

- Wong, George S. K.; Zhu, Shi-ming (1995). "Speed of sound in seawater as a function of salinity, temperature, and pressure". The Journal of the Acoustical Society of America. 97 (3): 1732. doi:10.1121/1.413048

- Murdin, Paul (25 December 2008). Full Meridian of Glory: Perilous Adventures in the Competition to Measure the Earth. Springer Science & Business Media. pp. 35–36. ISBN 9780387755342

- Robinson, Stephen (22 Sep 2005). "Technical Guides - Speed of Sound in Sea-Water". National Physical Laboratory. Retrieved 7 December 2016

- Stansfeld, Stephen A.; Matheson, Mark P. (2003-12-01). "Noise pollution: non-auditory effects on health". British Medical Bulletin. 68 (1): 243–257. ISSN 0007-1420. PMID 14757721. doi:10.1093/bmb/ldg033

- Kenneth V., Mackenzie (1981). "Discussion of sea-water sound-speed determinations". Journal of the Acoustical Society of America. 70 (3): 801–806. Bibcode:1981ASAJ...70..801M. doi:10.1121/1.386919

- Cozens, Peter (2006). The Darkest Days of the War: the Battles of Iuka and Corinth. Chapel Hill: The University of North Carolina Press. ISBN 0-8078-5783-1

- "EHP – Environmental Noise Pollution in the United States: Developing an Effective Public Health Response". ehp.niehs.nih.gov. Retrieved 2016-06-20

- Del Grosso, V. A. (1974). "New equation for speed of sound in natural waters (with comparisons to other equations)". Journal of the Acoustical Society of America. 56 (4): 1084–1091. Bibcode:1974ASAJ...56.1084D. doi:10.1121/1.1903388

- Dixon Ward, W. (1965). "Psychoacoustics". In Audiometry: Principles and Practices, edited by Aram Glorig, 55. Baltimore: Williams & Wilkins Co. Reprinted, Huntington, N.Y.: R. E. Krieger Pub. Co., 1977. ISBN 0-88275-604-4

- Murphy, William; Tak, SangWoo (2009-11-24). "Workplace Hearing Loss". Centers for Disease Control and Prevention. Retrieved 2016-06-15

- Passchier-Vermeer, W; Passchier, W F (2000-03-01). "Noise exposure and public health.". Environmental Health Perspectives. 108 (Suppl 1): 123–131. ISSN 0091-6765. PMC 1637786. PMID 10698728. doi:10.2307/3454637

- Grey, John M. (1977). "Multidimensional Perceptual Scaling of Musical Timbres". The Journal of the Acoustical Society of America 61(5):1270–77. doi:10.1121/1.381428

- Samson, Jim (1977). Music in Transition: A Study of Tonal Expansion and Atonality, 1900-1920. New York: W. W. Norton & Company. ISBN 0-393-02193-9

- "Standard Summary Project Fiche: Implementation Capacity for Environmental Noise Directive" (PDF). European Commission. Retrieved June 16, 2016

- Pijanowski, Bryan C.; Villanueva-Rivera, Luis J.; Dumyahn, Sarah L.; Farina, Almo; Krause, Bernie; Napoletano, Brian M.; Gage, Stuart H.; Pieretti, Nadia (March 2011). "Soundscape Ecology: The Science of Sound in the Landscape". BioScience. 61 (3): 203–216. doi:10.1525/bio.2011.61.3.6

- Wessel, David (1979). "Low Dimensional Control of Musical Timbre". Computer Music Journal 3:45–52. Rewritten version, 1999, as "Timbre Space as a Musical Control Structure"

- Elliot, Barry J. (2002). Designing a Structured Cabling System to ISO 11801 2nd Edition. Cambridge: Woodhead Publishing. p. 80. ISBN 978-0-8247-4130-3

- Mary Florentine (March 2003), It's not recruitment-gasp!! It's softness imperception, 56 (3), Hearing Journal, pp. 10, 12, 14, 15, doi:10.1097/01.HJ.0000293012.17887.b4

2

Understanding Audio Frequency

Audio frequency is the frequency that is sensed by the human ear. The unit used for audio frequency is hertz. This chapter has been carefully written to provide an easy understanding of the varied facets of audio frequency.

Audio Frequency

An audio frequency (abbreviation: AF) or audible frequency is characterized as a periodic vibration whose frequency is audible to the average human. The SI unit of audio frequency is the hertz (Hz). It is the property of sound that most determines pitch.

The generally accepted standard range of audible frequencies is 20 to 20,000 Hz, although the range of frequencies individuals hear is greatly influenced by environmental factors. Frequencies below 20 Hz are generally felt rather than heard, assuming the amplitude of the vibration is great enough. Frequencies above 20,000 Hz can sometimes be sensed by young people. High frequencies are the first to be affected by hearing loss due to age and/or prolonged exposure to very loud noises.

Hertz

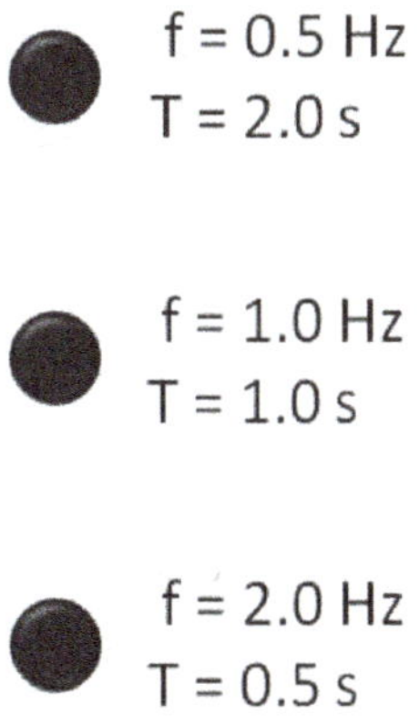

$f = 0.5$ Hz
$T = 2.0$ s

$f = 1.0$ Hz
$T = 1.0$ s

$f = 2.0$ Hz
$T = 0.5$ s

Top to bottom: Lights flashing at frequencies f = 0.5 Hz (hertz), 1.0 Hz and 2.0 Hz, i.e. at 0.5, 1.0 and 2.0 flashes per second, respectively. The time between each flash – the period T – is given by 1/f (the reciprocal of f), i.e. 2, 1 and 0.5 seconds, respectively.

The hertz (symbol Hz) is the unit of frequency in the International System of Units (SI) and is defined as one cycle per second. It is named for Heinrich Rudolf Hertz, the first person to provide conclusive proof of the existence of electromagnetic waves. Hertz are commonly expressed in multiples kilohertz (10^3 Hz, symbol kHz), megahertz (10^6 Hz, MHz), gigahertz (10^9 Hz, GHz), and terahertz (10^{12} Hz, THz). Kilo equals one thousand, mega equals one million, giga equals one billion and tera equals one trillion.

Some of the unit's most common uses are in the description of sine waves and musical tones, particularly those used in radio- and audio-related applications. It is also used to describe the speeds at which computers and other electronics are driven.

Definition

The hertz is equivalent to cycles per second, i.e., "1/second" or s^{-1}. The International Committee for Weights and Measures defined the second as "the duration of 9 192 631 770 periods of the radiation corresponding to the transition between the two hyperfine levels of the ground state of the caesium 133 atom" and then adds the obvious conclusion: "It follows that the hyperfine splitting in the ground state of the caesium 133 atom is exactly 9 192 631 770 hertz, v(hfs Cs) = 9 192 631 770 Hz."

In English, "hertz" is also used as the plural form. As an SI unit, Hz can be prefixed; commonly used multiples are kHz (kilohertz, 10^3 Hz), MHz (megahertz, 10^6 Hz), GHz (gigahertz, 10^9 Hz) and THz (terahertz, 10^{12} Hz). One hertz simply means «one cycle per second" (typically that which is being counted is a complete cycle); 100 Hz means "one hundred cycles per second", and so on. The unit may be applied to any periodic event—for example, a clock might be said to tick at 1 Hz, or a human heart might be said to beat at 1.2 Hz. The occurrence rate of aperiodic or stochastic events is expressed in reciprocal second or inverse second (1/s or s^{-1}) in general or, in the specific case of radioactive decay, in becquerels. Whereas 1 Hz is 1 cycle per second, 1 Bq is 1 aperiodic radionuclide event per second.

Even though angular velocity, angular frequency and the unit hertz all have the dimension 1/s, angular velocity and angular frequency are not expressed in hertz, but rather in an appropriate angular unit such as radians per second. Thus a disc rotating at 60 revolutions per minute (rpm) is said to be rotating at either 2π rad/s *or* 1 Hz, where the former measures the angular velocity and the latter reflects the number of *complete* revolutions per second. The conversion between a frequency *f* measured in hertz and an angular velocity ω measured in radians per second is

$$\omega = 2\pi f \ \text{ and } \ f = \frac{\omega}{2\pi}.$$

This SI unit is named after Heinrich Hertz. As with every International System of Units (SI) unit named for a person, the first letter of its symbol is upper case (Hz). However, when an SI unit is spelled out in English, it should always begin with a lower case letter (*hertz*)—except in a situation where *any* word in that position would be capitalized, such as at the beginning of a sentence or in material using title case. "Degree Celsius" conforms to this rule because the "d" is lowercase.—based on *The International System of Units*.

History

The hertz is named after the German physicist Heinrich Hertz (1857–1894), who made important scientific contributions to the study of electromagnetism. The name was established by the International Electrotechnical Commission (IEC) in 1930. It was adopted by the General Conference on Weights and Measures (CGPM) (*Conférence générale des poids et mesures*) in 1960, replacing the previous name for the unit, *cycles per second* (cps), along with its related multiples, primarily *kilocycles per second* (kc/s) and *megacycles per second* (Mc/s), and occasionally *kilomegacycles per second* (kMc/s). The term *cycles per second* was largely replaced by *hertz* by the 1970s. One hobby magazine, *Electronics Illustrated*, declared their intention to stick with the traditional kc., Mc., etc. units.

Applications

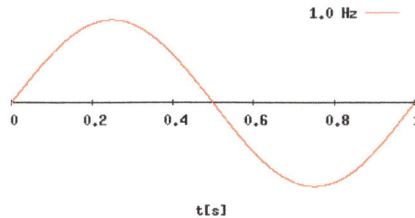

A sine wave with varying frequency

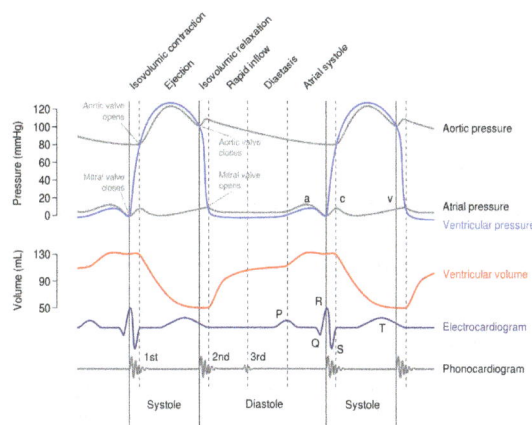

A heartbeat is an example of a non-sinusoidal periodic phenomenon
that may be analyzed in terms of frequency. Two cycles are illustrated.

Vibration

Sound is a traveling longitudinal wave which is an oscillation of pressure. Humans perceive frequency of sound waves as pitch. Each musical note corresponds to a particular frequency which can be measured in hertz. An infant's ear is able to perceive frequencies ranging from 20 Hz to 20,000 Hz; the average adult human can hear sounds between 20 Hz and 16,000 Hz. The range of ultrasound, infrasound and other physical vibrations such as molecular and atomic vibrations extends from a few femtoHz into the terahertz range and beyond.

Electromagnetic Radiation

Electromagnetic radiation is often described by its frequency—the number of oscillations of the perpendicular electric and magnetic fields per second—expressed in hertz.

Radio frequency radiation is usually measured in kilohertz (kHz), megahertz (MHz), or gigahertz (GHz). Light is electromagnetic radiation that is even higher in frequency, and has frequencies in the range of tens (infrared) to thousands (ultraviolet) of terahertz. Electromagnetic radiation with frequencies in the low terahertz range, (intermediate between those of the highest normally usable radio frequencies and long-wave infrared light), is often called terahertz radiation. Even higher frequencies exist, such as that of gamma rays, which can be measured in exahertz. (For historical reasons, the frequencies of light and higher frequency electromagnetic radiation are more commonly specified in terms of their wavelengths or photon energies: for a more detailed treatment of this and the above frequency ranges.)

Computers

In computers, most central processing units (CPU) are labeled in terms of their clock rate expressed in megahertz or gigahertz (10^6 or 10^9 hertz, respectively). This number refers to the frequency of the CPU's master clock signal ("clock rate"). This signal is a square wave, which is an electrical voltage that switches between low and high values at regular intervals. Hertz has become the primary unit of measurement accepted by the general populace to determine the performance of a CPU, but many experts have criticized this approach, which they claim is an easily manipulable benchmark as some processors use multiple clock periods to perform a single operation, while others can perform multiple operations in a single cycle. For home-based personal computers, CPU clock speeds have ranged from approximately 1 MHz in the late 1970s (Atari, Commodore, Apple computers) to up to 6 GHz in IBM POWER processors.

Various computer buses, such as the front-side bus connecting the CPU and northbridge, also operate at various frequencies in the megahertz range.

The "speed" or the more correct term bandwidth of networks, including wireless, such as Wi-Fi, 3G, 4G, LTE, etc. is not affected by the myth described above. In general, higher frequency allows for higher possible bandwidth or bit rates, and lowers the time it takes to download large files. In general, however, higher frequencies do not pass through walls, making lower frequencies an advantage indoors. Most common equipment reports the highest frequency available, but usually also works at lower ones. Bandwidth is often equated with "speed", while latency ("lag"), is the perceived problem (and is a measure distinct from bandwidth), while higher bandwidth has an influence on latency.

SI Multiples

SI multiples for hertz (Hz)					
Submultiples			**Multiples**		
Value	**SI symbol**	**Name**	**Value**	**SI symbol**	**Name**
10^{-1} Hz	dHz	decihertz	10^1 Hz	daHz	decahertz
10^{-2} Hz	cHz	centihertz	10^2 Hz	hHz	hectohertz
10^{-3} Hz	mHz	millihertz	10^3 Hz	**kHz**	**kilohertz**
10^{-6} Hz	μHz	microhertz	10^6 Hz	**MHz**	**megahertz**
10^{-9} Hz	nHz	nanohertz	10^9 Hz	**GHz**	**gigahertz**
10^{-12} Hz	pHz	picohertz	10^{12} Hz	**THz**	**terahertz**
10^{-15} Hz	fHz	femtohertz	10^{15} Hz	PHz	petahertz
10^{-18} Hz	aHz	attohertz	10^{18} Hz	EHz	exahertz
10^{-21} Hz	zHz	zeptohertz	10^{21} Hz	ZHz	zettahertz
10^{-24} Hz	yHz	yoctohertz	10^{24} Hz	YHz	yottahertz
Common prefixed units are in bold face.					

Higher frequencies than the International System of Units provides prefixes for are believed to occur naturally in the frequencies of the quantum-mechanical vibrations of high-energy, or, equivalently, massive particles, although these are not directly observable and must be inferred from their interactions with other phenomena. For practical reasons, these are typically not expressed in hertz, but in terms of the equivalent quantum energy, which is proportional to the frequency by the factor of Planck's constant.

Pitch (Music)

Pitch is a perceptual property of sounds that allows their ordering on a frequency-related scale, or more commonly, pitch is the quality that makes it possible to judge sounds as "higher" and "lower" in the sense associated with musical melodies. Pitch can be determined only in sounds that have a frequency that is clear and stable enough to distinguish from noise. Pitch is a major auditory attribute of musical tones, along with duration, loudness, and timbre.

Pitch may be quantified as a frequency, but pitch is not a purely objective physical property; it is a subjective psychoacoustical attribute of sound. Historically, the study of pitch and pitch perception has been a central problem in psychoacoustics, and has been instrumental in forming and testing theories of sound representation, processing, and perception in the auditory system.

Perception of Pitch

Pitch and Frequency

Pitch is an auditory sensation in which a listener assigns musical tones to relative positions on a musical scale based primarily on their perception of the frequency of vibration. Pitch is closely related to frequency, but the two are not equivalent. Frequency is an objective, scientific attribute that can be measured. Pitch is each person's *subjective perception* of a sound wave, which cannot be directly measured. However, this does not necessarily mean that most people won't agree on which notes are higher and lower.

Sound waves themselves do not have pitch, but their oscillations can be measured to obtain a frequency. It takes a sentient mind to map the internal quality of pitch. However, pitches are usually associated with, and thus quantified as frequencies in cycles per second, or hertz, by comparing sounds with pure tones, which have periodic, sinusoidal waveforms. Complex and aperiodic sound waves can often be assigned a pitch by this method.

According to the American National Standards Institute, pitch is the auditory attribute of sound according to which sounds can be ordered on a scale from low to high. Since pitch is such a close proxy for frequency, it is almost entirely determined by how quickly the sound wave is making the air vibrate and has almost nothing to do with the intensity, or amplitude, of the wave. That is, "high" pitch means very rapid oscillation, and "low" pitch corresponds to slower oscillation. Despite that, the idiom relating vertical height to sound pitch is shared by most languages. At least in English, it is just one of many deep conceptual metaphors that involve up/down. The exact etymological history of the musical sense of high and low pitch is still unclear. There is evidence that humans do actually perceive that the source of a sound is slightly higher or lower in vertical space when the sound frequency is increased or reduced.

In most cases, the pitch of complex sounds such as speech and musical notes corresponds very nearly to the repetition rate of periodic or nearly-periodic sounds, or to the reciprocal of the time interval between repeating similar events in the sound waveform.

The pitch of complex tones can be ambiguous, meaning that two or more different pitches can be perceived, depending upon the observer. When the actual fundamental frequency can be precisely determined through physical measurement, it may differ from the perceived pitch because of overtones, also known as upper partials, harmonic or otherwise. A complex tone composed of two sine waves of 1000 and 1200 Hz may sometimes be heard as up to three pitches: two spectral pitches at 1000 and 1200 Hz, derived from the physical frequencies of the pure tones, and the combination tone at 200 Hz, corresponding to the repetition rate of the waveform. In a situation like this, the percept at 200 Hz is commonly referred to as the missing fundamental, which is often the greatest common divisor of the frequencies present.

Pitch depends to a lesser degree on the sound pressure level (loudness, volume) of the tone, especially at frequencies below 1,000 Hz and above 2,000 Hz. The pitch of lower tones gets lower as sound pressure increases. For instance, a tone of 200 Hz that is very loud seems one semitone lower in pitch than if it is just barely audible. Above 2,000 Hz, the pitch gets higher as the sound gets louder.

Theories of Pitch Perception

Theories of pitch perception try to explain how the physical sound and specific physiology of the auditory system work together to yield the experience of pitch. In general, pitch perception theories can be divided into place coding and temporal coding. Place theory holds that the perception of pitch is determined by the place of maximum excitation on the basilar membrane.

A place code, taking advantage of the tonotopy in the auditory system, must be in effect for the perception of high frequencies, since neurons have an upper limit on how fast they can phase-lock their action potentials. However, a purely place-based theory cannot account for the accuracy of pitch perception in the low and middle frequency ranges.

Temporal theories offer an alternative that appeals to the temporal structure of action potentials, mostly the phase-locking and mode-locking of action potentials to frequencies in a stimulus. The precise way this temporal structure helps code for pitch at higher levels is still debated, but the processing seems to be based on an autocorrelation of action potentials in the auditory nerve. However, it has long been noted that a neural mechanism that may accomplish a delay—a necessary operation of a true autocorrelation—has not been found. At least one model shows that a temporal delay is unnecessary to produce an autocorrelation model of pitch perception, appealing to phase shifts between cochlear filters; however, earlier work has shown that certain sounds with a prominent peak in their autocorrelation function do not elicit a corresponding pitch percept, and that certain sounds without a peak in their autocorrelation function nevertheless elicit a pitch. To be a more complete model, autocorrelation must therefore apply to signals that represent the output of the cochlea, as via auditory-nerve interspike-interval histograms. Some theories of pitch perception hold that pitch has inherent octave ambiguities, and therefore is best decomposed into a pitch *chroma*, a periodic value around the octave, like the note names in western music—and a pitch *height*, which may be ambiguous, that indicates the octave the pitch is in.

Just-noticeable Difference

The *just-noticeable difference (jnd)* (the threshold at which a change is perceived) depends on the tone's frequency content. Below 500 Hz, the jnd is about 3 Hz for sine waves, and 1 Hz for complex tones; above 1000 Hz, the jnd for sine waves is about 0.6% (about 10 cents). The jnd is typically tested by playing two tones in quick succession with the listener asked if there was a difference in their pitches. The jnd becomes smaller if the two tones are played simultaneously as the listener is then able to discern beat frequencies. The total number of perceptible pitch steps in the range of human hearing is about 1,400; the total number of notes in the equal-tempered scale, from 16 to 16,000 Hz, is 120.

Aural Illusions

The relative perception of pitch can be fooled, resulting in *aural illusions*. There are several of these, such as the tritone paradox, but most notably the Shepard scale, where a continuous or discrete sequence of specially formed tones can be made to sound as if the sequence continues ascending or descending forever.

Definite and Indefinite Pitch

Not all musical instruments make notes with a clear pitch. The unpitched percussion instrument (a class of percussion instrument) does not produce particular pitches. A sound or note of definite pitch is one where a listener can possibly (or relatively easily) discern the pitch. Sounds with definite pitch have harmonic frequency spectra or close to harmonic spectra.

A sound generated on any instrument produces many modes of vibration that occur simultaneously. A listener hears numerous frequencies at once. The vibration with the lowest frequency is called the *fundamental frequency*; the other frequencies are *overtones*. *Harmonics* are an important class of overtones with frequencies that are integer multiples of the fundamental. Whether or not the higher frequencies are integer multiples, they are collectively called the partials, referring to the different parts that make up the total spectrum.

A sound or note of indefinite pitch is one that a listener finds impossible or relatively difficult to identify as to pitch. Sounds with indefinite pitch do not have harmonic spectra or have altered harmonic spectra—a characteristic known as inharmonicity.

It is still possible for two sounds of indefinite pitch to clearly be higher or lower than one another. For instance, a snare drum sounds higher pitched than a bass drum though both have indefinite pitch, because its sound contains higher frequencies. In other words, it is possible and often easy to roughly discern the relative pitches of two sounds of indefinite pitch, but sounds of indefinite pitch do not neatly correspond to any specific pitch. A special type of pitch often occurs in free nature when sound reaches the ear of an observer directly from the source, and also after reflecting off a sound-reflecting surface. This phenomenon is called *repetition pitch*, because the addition of a true repetition of the original sound to itself is the basic prerequisite.

Pitch Standards and Standard Pitch

A pitch standard (also Concert pitch) is the conventional pitch reference a group of musical instruments are tuned to for a performance. Concert pitch may vary from ensemble to ensemble, and has varied widely over musical history.

Standard pitch is a more widely accepted convention. The A above middle C is usually set at 440 Hz (often written as "A = 440 Hz" or sometimes "A440"), although other frequencies, such as 442 Hz, are also often used as variants. Another standard pitch, the so-called *Baroque pitch*, has been set in the 20th century as A = 415 Hz—approximately an equal-tempered semitone lower than A440 to facilitate transposition.

Transposing instruments have their origin in the variety of pitch standards. In modern times, they conventionally have their parts transposed into different keys from voices and other instruments (and even from each other). As a result, musicians need a way to refer to a particular pitch in an unambiguous manner when talking to each other.

For example, the most common type of clarinet or trumpet, when playing a note written in their part as C, sounds a pitch that is called B♭ on a non-transposing instrument like a violin (which indicates that at one time these wind instruments played at a standard pitch a tone lower than violin pitch). To refer to that pitch unambiguously, a musician calls it *concert B♭*, meaning, "...the pitch that someone playing a non-transposing instrument like a violin calls B♭."

Labeling Pitches

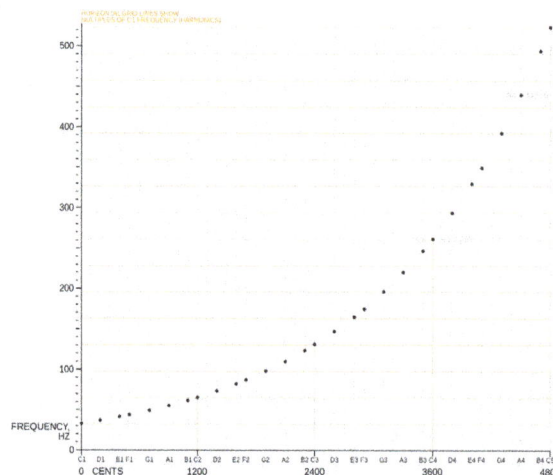

Note frequencies, four-octave C major diatonic scale, starting with C_1.

Pitches are labeled using:

- Letters, as in Helmholtz pitch notation

- A combination of letters and numbers—as in scientific pitch notation, where notes are labelled upwards from C_0, the 16 Hz C

- Numbers that represent the frequency in hertz (Hz), the number of cycles per second

For example, one might refer to the A above middle C as *a'*, A_4, or *440 Hz*. In standard Western equal temperament, the notion of pitch is insensitive to "spelling": the description "G_4 double sharp" refers to the same pitch as A_{4}; in other temperaments, these may be distinct pitches. Human perception of musical intervals is approximately logarithmic with respect to fundamental frequency: the perceived interval between the pitches "A220" and "A440" is the same as the perceived interval between the pitches *A440* and *A880*. Motivated by this logarithmic perception, music theorists sometimes represent pitches using a numerical scale based on the logarithm of fundamental frequency. For example, one can adopt the widely used MIDI standard to map fundamental frequency, f, to a real number, p, as follows

$$p = 69 + 12 \times \log_2\left(\frac{f}{440 \text{ Hz}}\right)$$

This creates a linear pitch space in which octaves have size 12, semitones (the distance between adjacent keys on the piano keyboard) have size 1, and A440 is assigned the number 69. Distance in this space corresponds to musical intervals as understood by musicians. An equal-tempered semitone is subdivided into 100 cents. The system is flexible enough to include "microtones" not found on standard piano keyboards. For example, the pitch halfway between C (60) and C♯ (61) can be labeled 60.5.

The following table shows frequencies in Hz for notes in various octaves, named according to the "German method" of octave nomenclature:

Note	Contra	Great	Small	One-lined	Two-lined	Three-lined
A	55.00	110.00	220.00	440.00	880.00	1760.00
A♯/B♭	58.27	116.54	233.08	466.16	932.33	1864.66
B/C♭	61.74	123.47	246.94	493.88	987.77	1975.53
B♯/C	65.41	130.81	261.63	523.25	1046.50	2093.00
C♯/D♭	69.30	138.59	277.18	554.37	1108.73	2217.46
D	73.42	146.83	293.66	587.33	1174.66	2349.32
D♯/E♭	77.78	155.56	311.13	622.25	1244.51	2489.02
E/F♭	82.41	164.81	329.63	659.26	1318.51	2637.02
E♯/F	87.31	174.61	349.23	698.46	1396.91	2793.83
F♯/G♭	92.50	185.00	369.99	739.99	1479.98	2959.96
G	98.00	196.00	392.00	783.99	1567.99	3135.96
G♯/A♭	103.83	207.65	415.30	830.61	1661.22	3322.44

Scales

The relative pitches of individual notes in a scale may be determined by one of a number of tuning systems. In the west, the twelve-note chromatic scale is the most common method of organization, with equal temperament now the most widely used method of tuning that scale. In it, the pitch ratio between any two successive notes of the scale is exactly the twelfth root of two (or about 1.05946). In well-tempered systems (as used in the time of Johann Sebastian Bach, for example), different methods of musical tuning were used. Almost all of these systems have one interval in common, the octave, where the pitch of one note is double the frequency of another. For example, if the A above middle C is 440 Hz, the A an octave above that is 880 Hz (info).

Other Musical Meanings of Pitch

In atonal, twelve tone, or musical set theory a "pitch" is a specific frequency while a pitch class is all the octaves of a frequency. In many analytic discussions of atonal and post-tonal music, pitches are named with integers because of octave and enharmonic equivalency (for example, in a serial system, C♯ and D♭ are considered the same pitch, while C_4 and C_5 are functionally the same, one octave apart).

Discrete pitches, rather than continuously variable pitches, are virtually universal, with exceptions including "tumbling strains" and "indeterminate-pitch chants". Gliding pitches are used in most cultures, but are related to the discrete pitches they reference or embellish.

Scale (Music)

In music theory, a scale is any set of musical notes ordered by fundamental frequency or pitch. A scale ordered by increasing pitch is an ascending scale, and a scale ordered by decreasing pitch is a descending scale. Some scales contain different pitches when ascending than when descending. For example, the Melodic minor scale.

Often, especially in the context of the common practice period, most or all of the melody and harmony of a musical work is built using the notes of a single scale, which can be conveniently represented on a staff with a standard key signature.

Due to the principle of octave equivalence, scales are generally considered to span a single octave, with higher or lower octaves simply repeating the pattern. A musical scale represents a division of the octave space into a certain number of scale steps, a scale step being the recognizable distance (or interval) between two successive notes of the scale. However, there is no need for scale steps to be equal within any scale and, particularly as demonstrated by microtonal music, there is no limit to how many notes can be injected within any given musical interval.

A measure of the width of each scale step provides a method to classify scales. For instance, in a chromatic scale each scale step represents a semitone interval, while a major scale is defined by the interval pattern T–T–S–T–T–T–S, where T stands for whole tone (an interval spanning two semitones), and S stands for semitone. Based on their interval patterns, scales are put into categories including diatonic, chromatic, major, minor, and others.

A specific scale is defined by its characteristic interval pattern and by a special note, known as its first degree (or tonic). The tonic of a scale is the note selected as the beginning of the octave, and therefore as the beginning of the adopted interval pattern. Typically, the name of the scale specifies both its tonic and its interval pattern. For example, C major indicates a major scale with a C tonic.

Background

Scales, Steps, and Intervals

Scales are typically listed from low to high pitch. Most scales are *octave-repeating*, meaning their pattern of notes is the same in every octave (the Bohlen–Pierce scale is one exception). An octave-repeating scale can be represented as a circular arrangement of pitch classes, ordered by increasing (or decreasing) pitch class. For instance, the increasing C major scale is C–D–E–F–G–A–B–[C], with the bracket indicating that the last note is an octave higher than the first note, and the decreasing C major scale is C–B–A–G–F–E–D–[C], with the bracket indicating an octave lower than the first note in the scale.

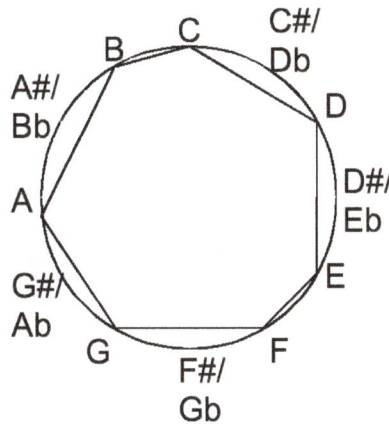

Diatonic scale in the chromatic circle

The distance between two successive notes in a scale is called a scale step.

The notes of a scale are numbered by their steps from the root of the scale. For example, in a C major scale the first note is C, the second D, the third E and so on. Two notes can also be numbered in relation to each other: C and E create an interval of a third (in this case a major third); D and F also create a third (in this case a minor third).

Pitch

A single scale can be manifested at many different pitch levels. For example, a C major scale can be started at C4 (middle C) and ascending an octave to C5; or it could be started at C6, ascending an octave to C7. As long as all the notes can be played, the octave they take on can be altered.

Types of Scale

Scales may be described according to the intervals they contain:

- for example: diatonic, chromatic, whole tone

or by the number of different pitch classes they contain:

- Octatonic (8 notes per octave): used in jazz and modern classical music

- Heptatonic (7 notes per octave): the most common modern Western scale

- Hexatonic (6 notes per octave): common in Western folk music

- Pentatonic (5 notes per octave): the anhemitonic form (lacking semitones) is common in folk music, especially in oriental music; also known as the "black note" scale

- Tetratonic (4 notes), tritonic (3 notes), and ditonic (2 notes): generally limited to prehistoric ("primitive") music

- Monotonic (1 note): limited use in liturgy, and for effect in modern art music

"The number of the notes that make up a scale as well as the quality of the intervals between successive notes of the scale help to give the music of a culture area its peculiar sound quality." "The pitch distances or intervals among the notes of a scale tell us more about the sound of the music than does the mere number of tones."

Harmonic Content

The notes of a scale form intervals with each of the other notes of the chord in combination. A 5-note scale has 10 of these harmonic intervals, a 6-note scale has 15, a 7-note scale has 21, an 8-note scale has 28. Though the scale is not a chord, and might never be heard more than one note at a time, still the absence, presence, and placement of certain key intervals plays a large part in the sound of the scale, the natural movement of melody within the scale, and the selection of chords taken naturally from the scale.

A musical scale that contains tritones is called tritonic (though the expression is also used for any scale with just three notes per octave, whether or not it includes a tritone), and one without tritones is *atritonic*. A scale or chord that contains semitones is called hemitonic, and without semitones is anhemitonic. The significance of these categories lies in their bases of semitones and tritones being the severest of dissonances, which is often desirable to avoid. Most scales used across the planet are anhemitonic.

Scales in Composition

Scales can be abstracted from performance or composition. They are also often used precompositionally to guide or limit a composition. Explicit instruction in scales has been part of compositional training for many centuries. One or more scales may be used in a composition, such as in Claude Debussy's *L'Isle Joyeuse*. To the right, the first scale is a whole tone scale, while the second and third scales are diatonic scales. All three are used in the opening pages of Debussy's piece.

Western Music

Scales in traditional Western music generally consist of seven notes and repeat at the octave. Notes in the commonly used scales are separated by whole and half step intervals of *tones* and *semitones*. The harmonic minor scale includes a three-semitone step; the anhemitonic pentatonic includes two of those and no semitones.

Western music in the Medieval and Renaissance periods (1100–1600) tends to use the white-note diatonic scale C–D–E–F–G–A–B. Accidentals are rare, and somewhat unsystematically used, often to avoid the tritone.

Music of the common practice periods (1600–1900) uses three types of scale:

- The diatonic scale (seven notes)—this includes the major scale and the natural minor

- The melodic and harmonic minor scales (seven notes)

These scales are used in all of their transpositions. The music of this period introduces *modulation,* which involves systematic changes from one scale to another. Modulation occurs in relatively conventionalized ways. For example, major-mode pieces typically begin in a "tonic" diatonic scale and modulate to the "dominant" scale a fifth above.

In the 19th century (to a certain extent), but more in the 20th century, additional types of scales were explored:

- The chromatic scale (twelve notes)

- The whole tone scale (six notes)

- The pentatonic scale (five notes)

- The octatonic or diminished scales (eight notes)

A large variety of other scales exists, some of the more common being:

- The Phrygian dominant scale (a mode of the harmonic minor scale)

- The Arabic scales

- The Hungarian minor scale

- The Byzantine Music scales (called echoi)

- The Persian scale

Scales such as the pentatonic scale may be considered *gapped* relative to the diatonic scale. An *auxiliary scale* is a scale other than the primary or original scale.

Note Names

In many musical circumstances, a specific note of the scale is chosen as the tonic—the central and most stable note of the scale, also known as the root note. In Western tonal music, simple songs or pieces typically start and end on the tonic note. Relative to a choice of a certain tonic, the notes of a scale are often labeled with numbers recording how many scale steps above the tonic they are. For example, the notes of the C major scale (C, D, E, F, G, A, B) can be labeled {1, 2, 3, 4, 5, 6, 7}, reflecting the choice of C as tonic. The expression scale degree refers to these numerical labels. Such labeling requires the choice of a "first" note; hence scale-degree labels are not intrinsic to the scale itself, but rather to its modes. For example, if we choose A as tonic, then we can label the notes of the C major scale using A = 1, B = 2, C = 3, and so on. When we do so, we create a new scale called the A minor scale.

The scale degrees of a heptatonic (7-note) scale can also be named using the terms tonic, supertonic, mediant, subdominant, dominant, submediant, subtonic. If the subtonic is a semitone away from the tonic, then it is usually called the leading-tone (or leading-note); otherwise the leading-tone refers to the raised subtonic. Also commonly used is the (movable do) solfège naming convention in which each scale degree is denoted by a syllable. In the major scale, the solfege syllables are: Do, Re, Mi, Fa, So (or Sol), La, Ti (or Si), Do (or Ut).

In naming the notes of a scale, it is customary that each scale degree be assigned its own letter name: for example, the A major scale is written A–B–C♯–D–E–F♯–G♯ rather than A–B–D♭–D–E–E𝄪–G♯. However, it is impossible to do this in scales that contain more than seven notes, at least in the English-language nomenclature system.

Scales may also be identified by using a binary system of twelve zeros or ones to represent each of the twelve notes of a chromatic scale. It is assumed that the scale is tuned using 12-tone equal temperament (so that, for instance, C♯ is the same as D♭), and that the tonic is in the leftmost position. For example the binary number 101011010101, equivalent to the decimal number 2773, would represent any major scale (such as C–D–E–F–G–A–B). This system includes scales from 100000000000 (2048) to 111111111111 (4095), providing a total of 2048 possible species, but only 352 unique scales containing from 1 to 12 notes.

A more mathematically flexible schema for binary representation, espoused by theorist Ian Ring in his study of musical scales, has bits representing ascending tones, with the smaller bits representing lower pitches and higher bits representing higher pitches. This causes the bits to appear reversed compared with Duncan's system. Under this schema, a major scale is 101010110101 = (2741), a single root tone is 1, and a chromatic scale is still 111111111111 (4095). All the tone possibilities that begin with a root tone, and are thus scale-like, are odd-numbered.

Scales may also be shown as semitones from the tonic. For instance, 0 2 4 5 7 9 11 denotes any major scale such as C–D–E–F–G–A–B, in which the first degree is, obviously, 0 semitones from the tonic (and therefore coincides with it), the second is 2 semitones from the tonic, the third is 4 semitones from the tonic, and so on. Again, this implies that the notes are drawn from a chromatic scale tuned with 12-tone equal temperament. For some fretted string instruments, such as the guitar and the bass guitar, scales can be notated in tabulature, an approach which indicates the fret number and string upon which each scale degree is played.

Transposition and Modulation

Composers transform musical patterns by moving every note in the pattern by a constant number of scale steps: thus, in the C major scale, the pattern C–D–E might be shifted up, or transposed, a single scale step to become D–E–F. This process is called "scalar transposition" or "shifting to a new key" and can often be found in musical sequences and patterns. Since the steps of a scale can have various sizes, this process introduces subtle melodic and harmonic variation into the music. In Western tonal music, the simplest and most common type of modulation (or changing keys) is to shift from one major key to another key built on the first key's fifth (or dominant) scale degree. In the key of C Major, this would involve moving to the key of G Major (which uses an F#). Composers also often modulate to other related keys. In some Romantic music era pieces and contemporary music, composers modulate to "remote keys" that are not related to or close to the tonic. An example of a remote modulation would be taking a song that begins in C Major and modulating (changing keys) to F# Major. This variation from transposition and modulation is what gives scalar music much of its complexity.

Jazz and Blues

Through the introduction of blue notes, jazz and blues employ scale intervals smaller than a semitone. The blue note is an interval that is technically neither major nor minor but "in the middle", giving it a characteristic flavour. A regular piano cannot play blue notes, but with electric guitar, saxophone, trombone and trumpet, performers can "bend" notes a fraction of a tone sharp or flat to create blue notes For instance, in the key of E, the blue note would be either a note between G and G♯ or a note moving between both. In blues. a pentatonic scale is often used. In jazz many different modes and scales are used, often within the same piece of music. Chromatic scales are common, especially in modern jazz.

Non-western Scales

In Western music, scale notes are often separated by equally tempered tones or semitones, creating 12 notes per octave. Many other musical traditions use scales that include other intervals or a different number of pitches. These scales originate within the derivation of the harmonic series. Musical intervals are complementary values of the harmonic overtones series. Many musical scales in the world are based on this system, except most of the musical scales from Indonesia and the Indochina Peninsulae, which are based on inharmonic resonance of the dominant metalophone and xylophone instruments. A common scale in Eastern music is the pentatonic scale, consisting of five notes. The Middle Eastern Hejaz scale has some intervals of three semitones. Gamelan music uses a small variety of scales including Pélog and Sléndro, none including equally tempered nor harmonic intervals. Indian classical music uses a moveable seven-note scale. Indian Rāgas often use intervals smaller than a semitone. Arabic music maqamat may use quarter tone intervals. In both rāgas and maqamat, the distance between a note and an inflection (e.g., śruti) of that same note may be less than a semitone.

Microtonal Scales

The term *microtonal music* usually refers to music with roots in traditional Western music that uses non-standard scales or scale intervals. In the late 19th century, Mexican composer Julián Carrillo created microtonal scales that he called *Sonido 13*. The composer Harry Partch made custom musical instruments to play compositions based on 43-note scale system, and the American jazz vibraphonist Emil Richards experimented with such scales in his Microtonal Blues Band in the 1970s. Easley Blackwood wrote compositions in all equal-tempered scales from 13 to 24 notes. Erv Wilson introduced concepts such as Combination Product Sets (Hexany), Moments of Symmetry and golden horagrams, used by many modern composers. Microtonal scales are also used in traditional Indian Raga music, which uses a variety of modes not only as modes or scales, but also as defining elements of the song, or raga.

Fundamental Frequency

The fundamental frequency, often referred to simply as the fundamental, is defined as the lowest frequency of a periodic waveform. In music, the fundamental is the musical pitch of a note that is perceived as the lowest partial present. In terms of a superposition of sinusoids (e.g. Fourier series), the fundamental frequency is the lowest frequency sinusoidal in the sum. In some contexts, the fundamental is usually abbreviated as f_0 (or FF), indicating the lowest frequency counting from zero. In other contexts, it is more common to abbreviate it as f_1, the first harmonic. (The second harmonic is then $f_2 = 2 \cdot f_1$, etc. In this context, the zeroth harmonic would be 0 Hz.)

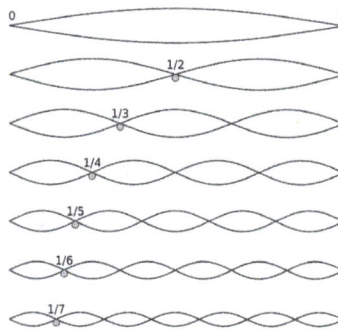

Vibration and standing waves in a string, The fundamental and the first six overtones

Explanation

All sinusoidal and many non-sinusoidal waveforms are periodic, which is to say they repeat exactly over time. The period of a waveform is the T for which the following equation is true:

$$x(t) = x(t+T) \text{ for all } t \in \mathbb{R}$$

Where $x(t)$ is the value of the waveform at t. This means that this equation and a definition of the waveforms values over any interval of length T is all that is required to describe the waveform completely.

Every waveform may be described using any multiple of this period. There exists a smallest period over which the function may be described completely and this period is the fundamental period. The fundamental frequency is defined as its reciprocal:

$$f_0 = \frac{1}{T}$$

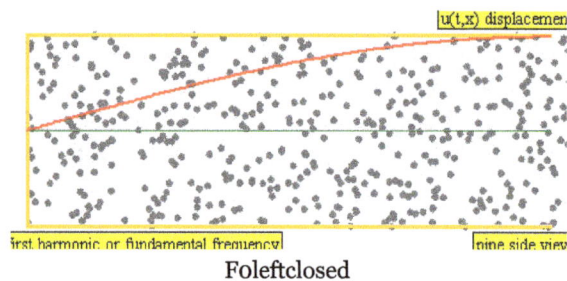

Foleftclosed

Since the period is measured in units of time, then the units for frequency are 1/time. When the time units are seconds, the frequency is in s^{-1}, also known as Hertz.

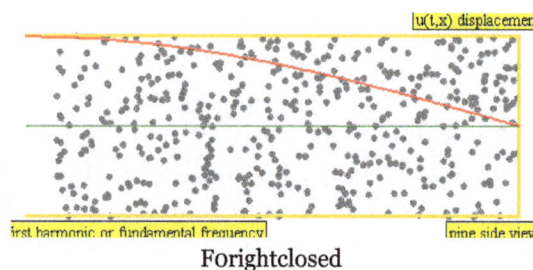

Forightclosed

For a tube of length L with one end closed and the other end open the wavelength of the fundamental harmonic is $4L$, as indicated by the first two figure. Hence,

$$\lambda_0 = 4L.$$

Therefore, using the relation

$$\lambda_0 = \frac{v}{f_0},$$

where v is the speed of the wave, we can find the fundamental frequency in terms of the speed of the wave and the length of the tube:

$$f_0 = \frac{v}{4L}.$$

Fobothclosed

Fobothopen

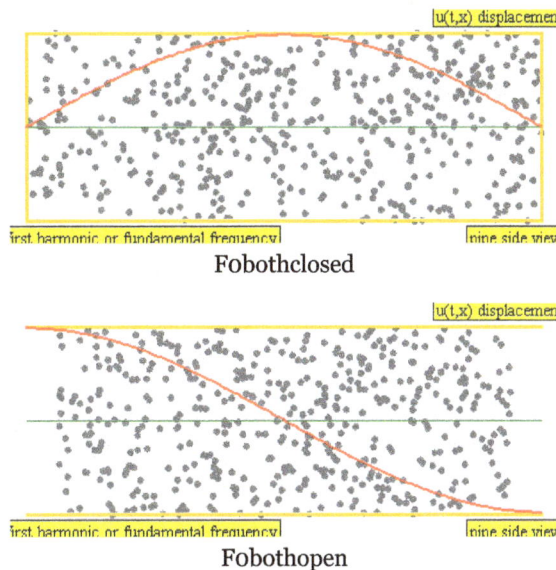

If the ends of the same tube are now both closed or both opened as in the last two animations, the wavelength of the fundamental harmonic becomes $2L$. By the same method as above, the fundamental frequency is found to be

$$f_0 = \frac{v}{2L}.$$

At 20° C (68° F) the speed of sound in air is 343 m/s (1129 ft/s). This speed is temperature dependent and increases at a rate of 0.6 m/s for each degree Celsius increase in temperature (1.1 ft/s for every increase of 1° F).

The velocity of a sound wave at different temperatures:-

- v = 343.2 m/s at 20° C

- v = 331.3 m/s at 0° C

In Music

In music, the fundamental is the musical pitch of a note that is perceived as the lowest partial present. The fundamental may be created by vibration over the full length of a string or air column, or a higher harmonic chosen by the player. The fundamental is one of the harmonics. A harmonic is any member of the harmonic series, an ideal set of frequencies that are positive integer multiples of a common fundamental frequency. The reason a fundamental is also considered a harmonic is because it is 1 times itself.

The fundamental is the frequency at which the entire wave vibrates. Overtones are other sinusoidal components present at frequencies above the fundamental. All of the frequency components that make up the total waveform, including the fundamental and the overtones, are called partials. Together they form the harmonic series. Overtones which are perfect integer multiples of the fundamental are called harmonics. When an overtone is near to being harmonic, but not exact, it is sometimes called a harmonic partial, although they are often referred to simply as harmonics. Sometimes overtones are created that are not anywhere near a harmonic, and are just called partials or inharmonic overtones.

The fundamental frequency is considered the *first harmonic* and the *first partial*. The numbering of the partials and harmonics is then usually the same; the second partial is the second harmonic, etc. But if there are inharmonic partials, the numbering no longer coincides. Overtones are numbered as they appear *above* the fundamental. So strictly speaking, the *first* overtone is the *second* partial (and usually the *second* harmonic). As this can result in confusion, only harmonics are usually referred to by their numbers, and overtones and partials are described by their relationships to those harmonics.

Mechanical Systems

Consider a spring, fixed at one end and having a mass attached to the other; this would be a single degree of freedom (SDoF) oscillator. Once set into motion it will oscillate at its natural frequency. For a single degree of freedom oscillator, a system in which the motion can be described by a single coordinate, the natural frequency depends on two system properties: mass and stiffness; (providing the system is undamped). The radian frequency, ω_n, can be found using the following equation:

$$\omega_n^2 = \frac{k}{m}$$

Where:

k = stiffness of the spring

m = mass

ω_n = radian frequency (radians per second)

From the radian frequency, the natural frequency, f_n, can be found by simply dividing ω_n by 2π. Without first finding the radian frequency, the natural frequency can be found directly using:

$$f_n = \frac{1}{2\pi} \sqrt{\frac{k}{m}}$$

Where:

f_n = natural frequency in hertz (cycles/second)

k = stiffness of the spring (Newtons/meter or N/m)

m = mass(kg)

while doing the modal analysis of structures and mechanical equipment, the frequency of 1st mode is called fundamental frequency.

Zero-based Numbering

Zero-based numbering or *index origin = 0* is a way of numbering in which the initial element of a sequence is assigned the index 0, rather than the index 1 as is typical in everyday *non-mathematical/non-programming* circumstances. Under zero-based numbering, the initial element is sometimes termed the *zeroth* element, rather than the *first* element; *zeroth* is a coined ordinal number corresponding to the number zero. In some cases, an object or value that does not (originally) belong to a given sequence, but which could be naturally placed before its initial element, may be termed the zeroth element. There is not wide agreement regarding the correctness of using zero as an ordinal (nor regarding the use of the term *zeroth*) as it creates ambiguity for all subsequent elements of the sequence when lacking context.

Numbering sequences starting at 0 is quite common in mathematics notation, in particular in combinatorics, though programming languages for mathematics usually index from 1. In computer science, array indices usually start at 0 in modern programming languages, so computer programmers might use *zeroth* in situations where others might use *first*, and so forth. In some mathematical contexts, zero-based numbering can be used without confusion, when ordinal forms have well established meaning with an obvious candidate to come before *first*; for instance a *zeroth derivative* of a function is the function itself, obtained by differentiating zero times. Such usage corresponds to naming an element not properly belonging to the sequence but preceding it: the zeroth derivative is not really a derivative at all. However, just as the *first derivative* precedes the *second derivative*, so also does the *zeroth derivative* (or the original function itself) precede the *first derivative*.

Computer Programming

The primary origin of using zero in enumerating data entities in computer programming lies in the available machine instructions, which allowed to "jump" in the linear sequence of instructions, given a specified condition. This is a crucial necessity to implement branching in instruction sequences. The main available conditions were "on zero" and "on negative" of the content in a specified register. One of the most frequently used instructions for implementing loops was a "decrement and jump if zero". That is, any data structure worth enumerating was dealt with in loops and zero was at hand to control these loops. All higher level language loop control structures (do-while, repeat-until, for-step-to, ...) are based on these elementary machine instructions and would not suggest the use of zero themselves to this extent.

Martin Richards, creator of the BCPL language (a precursor of C), designed arrays initiating at 0 as the natural position to start accessing the array contents in the language, since the value of a pointer p used as an address accesses the position p+0 in memory. Canadian systems analyst Mike Hoye asked Richards the reasons for choosing that convention. BCPL was first compiled for the IBM 7094; the language introduced no indirection lookups at run time, so the indirection optimization provided by these arrays was used at compile time. The optimization was nevertheless important, as batch processes in the system could be interrupted at any time to calculate yacht handicapping for the president of IBM's racing yacht.

E. Dijkstra later wrote a pertinent note *Why numbering should start at zero* in 1982, analyzing the possible designs of array indices by enclosing them in a chained inequality, combining sharp and standard inequalities to four possibilities, demonstrating that to his conviction zero-based arrays are best represented by non-overlapping index ranges, which start at zero, alluding to open, half-open and closed intervals as with the real numbers. Dijkstra's criteria for preferring this convention are in detail that it represents empty sequences in a more natural way ($a \le i < a$?) than closed "intervals" ($a \le i \le (a-1)$?), and that with half-open "intervals" of naturals, the length of a sub-sequence equals the upper minus the lower bound ($a \le i < b$ gives $(b-a)$ possible values for i, with a, b, i all naturals).

Usage in Programming Languages

This usage follows from design choices embedded in many influential programming languages, including C, Java, and Lisp. In these three, sequence types (C arrays, Java arrays and lists, and Lisp lists and vectors) are indexed beginning with the zero subscript. Particularly in C, where arrays are closely tied to pointer arithmetic, this makes for a simpler implementation: the subscript refers to an offset from the starting position of an array, so the first element has an offset of zero.

Referencing memory by an address and an offset is represented directly in computer hardware on virtually all computer architectures, so this design detail in C makes compilation easier, at the cost of some human factors. In this context using "zeroth" as an ordinal is not strictly correct, but a widespread habit in this profession. Other programming languages, such as Fortran or COBOL, have array subscripts starting with one, because they were meant as high-level programming languages, and as such they had to have a correspondence to the usual ordinal numbers. Some recent languages, such as Lua, have adopted the same convention for the same reason.

Zero is the lowest unsigned integer value, one of the most fundamental types in programming and hardware design. In computer science, zero is thus often used as the base case for many kinds of numerical recursion. Proofs and other sorts of mathematical reasoning in computer science often begin with zero. For these reasons, in computer science it is not unusual to number from zero rather than one.

Hackers and computer scientists often like to call the first chapter of a publication "Chapter 0", especially if it is of an introductory nature. One of the classic instances was in the First Edition of K&R. In recent years this trait has also been observed among many pure mathematicians, where many constructions are defined to be numbered from 0.

If an array is used to represent a cycle, it is convenient to obtain the index with a modulo function, which can result in zero.

Numerical Properties

With zero-based numbering, a range can be expressed as the half-open interval, [0,*n*), as opposed to the closed interval, [1,*n*]. Empty ranges, which often occur in algorithms, are tricky to express with a closed interval without resorting to obtuse conventions like [1,0]. Because of this property, zero-based indexing potentially reduces off-by-one and fencepost errors. On the other hand, the repeat count *n* is calculated in advance, making the use of counting from 0 to *n*−1 (inclusive) less intuitive. Some authors prefer one-based indexing as it corresponds more closely to how entities are indexed in other contexts.

Another property of this convention is in the use of modular arithmetic as implemented in modern computers. Usually, the modulo function maps any integer modulo *N* to one of the numbers 0, 1, 2, ..., *N* − 1, where *N* ≥ 1. Because of this, many formulas in algorithms (such as that for calculating hash table indices) can be elegantly expressed in code using the modulo operation when array indices start at zero.

Pointer operations can also be expressed more elegantly on a zero-based index due to the underlying address/offset logic mentioned above. To illustrate, suppose *a* is the memory address of the first element of an array, and *i* is the index of the desired element. To compute the address of the desired element, if the index numbers count from 1, the desired address is computed by this expression:

$$a + s \times (i - 1)$$

where *s* is the size of each element. In contrast, if the index numbers count from 0, the expression becomes:

$$a + s \times i$$

This simpler expression is more efficient to compute at run time.

Note, however, that a language wishing to index arrays from 1 could adopt the convention that every array address is represented by $a' = a - s$; that is, rather than using the address of the first array element, such a language would use the address of an "imaginary" element located immediately before the first actual element. The indexing expression for a 1-based index would be the following:

$$a' + s \times i$$

Hence, the efficiency benefit at run time of zero-based indexing is not inherent, but is an artifact of the decision to represent an array with the address of its first element rather than the address of the imaginary element preceding the array. However, the address of that imaginary element located immediately before the first actual element of the array could very well be the address of some other item in memory not related to the array.

This situation can lead to some confusion in terminology. In a zero-based indexing scheme, the first element is "element number zero"; likewise, the twelfth element is "element number eleven". Therefore, an analogy from the ordinal numbers to the quantity of objects numbered appears; the highest index of *n* objects will be *n* − 1 and referred to the *n*th element. For this reason, the first element is often referred to as the *zeroth* element to avoid confusion.

Science

In mathematics, many sequences of numbers or of polynomials are indexed by nonnegative integers, for example the Bernoulli numbers and the Bell numbers.

The *zeroth law of thermodynamics* was formulated after the first, second, and third laws, but considered more fundamental, thus its name.

In biology, an organism is said to have zero order intentionality if it shows "no intention of anything at all". This would include a situation where the organism's genetically predetermined phenotype results in a fitness benefit to itself, because it did not "intend" to express its genes. In the similar sense, a computer may be considered from this perspective a zero order intentional entity as it does not "intend" to express the code of the programs it runs.

In biological or medical experiments, initial measurements made before any experimental time has passed are said to be on the 0 day of the experiment.

In genomics, both 0-based and 1-based systems are used for genome coordinates.

Patient zero (or index case) is the initial patient in the population sample of an epidemiological investigation.

Other Fields

In the realm of fiction, Isaac Asimov eventually added a Zeroth Law to his Three Laws of Robotics, essentially making them four laws.

The year zero does not exist in the widely used Gregorian calendar or in its predecessor, the Julian calendar. Under those systems, the year 1 BC is followed by AD 1. However, there is a year zero in astronomical year numbering (where it coincides with the Julian year 1 BC) and in ISO 8601:2004 (where it coincides with the Gregorian year 1 BC) as well as in all Buddhist and Hindu calendars.

In many countries, the ground floor in buildings is considered as floor number 0 rather than as the "1st Floor", the naming convention usually found in the United States of America. This makes a consistent set with underground floors marked with negative numbers.

While the ordinal of 0 is rarely used outside communities closely connected to mathematics, physics, and computer science, there are a few instances in classical music. The composer Anton Bruckner regarded his early *Symphony in D minor* to be unworthy of including in the canon of his works, and he wrote 'gilt nicht' on the score and a circle with a crossbar, intending it to mean "invalid". But posthumously, this work came to be known as *Symphony No. 0 in D minor*, even though it was actually written after *Symphony No. 1 in C minor*. There is an even earlier *Symphony in F minor* of Bruckner's that is sometimes called *No. 00*. The Russian composer Alfred Schnittke also wrote a Symphony No. 0.

In some universities, including Oxford and Cambridge, "week 0" or occasionally "noughth week" refers to the week before the first week of lectures in a term. In Australia, some universities refer to this as "O Week", which serves as a pun on "orientation week". As a parallel, the introductory weeks at university educations in Sweden are generally called "nollning" (zeroing).

The United States Air Force starts basic training each Wednesday, and the first week (of eight) is considered to begin with the following Sunday. The four days before that Sunday are often referred to as "Zero Week."

24-hour clocks and the international standard ISO 8601 use 0 to denote the beginning of the day.

In London King's Cross, Uppsala, Yonago, Edinburgh Haymarket, Stockport and Cardiff the train stations have a platform 0.

Robert Crumb's drawings for the first issue of *Zap Comix* were stolen, so he drew a whole new issue which was published as issue 1. Later he re-inked his photocopies of the stolen artwork and published it as issue 0.

The ring road around Brussels is called R0. It was built after the ring road around Antwerp, but Brussels (being the capital city) was deemed deserving of a more basic number.

In Formula One, when a defending world champion does not compete in the following season, the number 1 is not assigned to any driver, but one driver of the world champion team will carry the number 0, and the other, number 2. This did happen both in 1993 and 1994 with Damon Hill carrying the number 0 in both seasons, as defending champion Nigel Mansell quit after 1992, and defending champion Alain Prost quit after 1993.

A chronological prequel of a series may be numbered as 0, such as *Ring 0: Birthday* or *Zork Zero*.

The Swiss Federal Railways number certain classes of rolling stock from zero, for example, Re 460 000 to 118.

References

- Cariani, P.A.; Delgutte, B. (September 1996). "Neural Correlates of the Pitch of Complex Tones. I. Pitch and Pitch Salience" (PDF). Journal of Neurophysiology. 76 (3): 1698–1716. PMID 8890286. Retrieved 13 November 2012

- Anssi Klapuri, "Introduction to Music Transcription", in Signal Processing Methods for Music Transcription, edited by Anssi Klapuri and Manuel Davy, 1–20 (New York: Springer, 2006): p. 8. ISBN 978-0-387-30667-4

- Heffner, Henry; Heffner, Rickye (January 2007). "Hearing Ranges of Laboratory Animals". American Association for Laboratory Animal Science. 46 (1): 20. Retrieved 19 September 2014

- Bitner-Glindzicz, M (2002). "Hereditary deafness and phenotyping in humans.". British medical bulletin. 63 (1): 73–94. PMID 12324385. doi:10.1093/bmb/63.1.73

- Plack, Christopher J.; Andrew J. Oxenham; Richard R. Fay, eds. (2005). Pitch: Neural Coding and Perception. Springer. ISBN 0-387-23472-1

- Dijkstra, Edsger Wybe (May 2, 2008). "Why numbering should start at zero (EWD 831)". E. W. Dijkstra Archive. University of Texas at Austin. Retrieved 2011-03-16

- Schwartz, David A.; Dale Purves (May 2004). "Pitch Is Determined by Naturally Occurring Periodic Sounds". Hearing Research. 194: 31–46. doi:10.1016/j.heares.2004.01.019

- Kaernbach, C.; Demany, L. (October 1998). "Psychophysical Evidence Against the Autocorrelation Theory of Auditory Temporal Processing". Journal of the Acoustical Society of America. 104 (4): 2298–2306. Bibcode:1998ASAJ..104.2298K. PMID 10491694. doi:10.1121/1.423742

- Birger Kollmeier; Thomas Brand & B. Meyer (2008). "Perception of Speech and Sound". In Jacob Benesty; M.

Mohan Sondhi & Yiteng Huang. Springer Handbook of Speech Processing. Springer. p. 65. ISBN 978-3-540-49125-5

- Cheveigné, A. de; Pressnitzer, D. (June 2006). "The Case of the Missing Delay Lines: Synthetic Delays Obtained by Cross-channel Phase Interaction" (PDF). Journal of the Acoustical Society of America. 119 (6): 3908–3918. Bibcode:2006ASAJ..119.3908D. PMID 16838534. doi:10.1121/1.2195291. Retrieved 13 November 2012

- Burns, E.M.; Viemeister, N. F. (October 1976). "Nonspectral Pitch". Journal of the Acoustical Society of America. 60 (4): 863–69. Bibcode:1976ASAJ...60..863B. doi:10.1121/1.381166

- Burns, Edward M. (1999). "Intervals, Scales, and Tuning", The Psychology of Music, second edition. Deutsch, Diana, ed. San Diego: Academic Press. ISBN 0-12-213564-4

- Pressnitzer, D.; Cheveigné, A. de; Winter, I.M. (January 2002). "Perceptual Pitch Shift for Sounds with Similar Waveform Autocorrelation". Acoustics Research Letters Online. 3 (1): 1–6. doi:10.1121/1.1416671

- John R. Pierce (2001). "Consonance and Scales". In Perry R. Cook. Music, Cognition, and Computerized Sound. MIT Press. ISBN 978-0-262-53190-0

- Brown, Jim (December 1978). "In Defense of Index Origin 0". ACM SIGAPL APL Quote Quad. 9 (2): 7. doi:10.1145/586050.586053

3

Equipments used in Audio Engineering

Devices that can modify, create and convert audio signals into an output are known as audio electronics. The other equipments discussed are analog-to-digital converter, loudspeaker, mixing console, tape recorder and music sequencer. The topics discussed in the chapter are of great importance to broaden the existing knowledge on the equipments used in audio engineering.

Audio Electronics

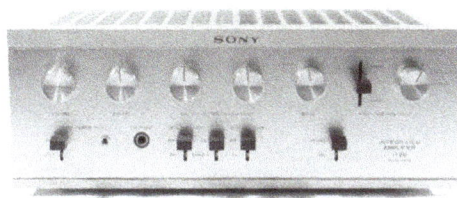

Audio amplifier

Audio electronics is the implementation of electronic circuit designs to perform conversions of sound/pressure wave signals to electrical signals, or vice versa. Electronic circuits considered a part of audio electronics may also be designed to achieve certain signal processing operations, in order to make particular alterations to the signal while it is in the electrical form. Additionally, audio signals can be created synthetically through the generation of electric signals from electronic devices. Audio Electronics were traditionally designed with analog electric circuit techniques until advances in digital technologies were developed. Moreover, digital signals are able to be manipulated by computer software much the same way audio electronic devices would, due to its compatible digital nature. Both analog and digital design formats are still used today, and the use of one or the other largely depends on the application. The following is a partial list of audio-related circuits/techniques/devices:

Basic Components

- Amplifiers
- Car audio
- Compressors
- Crossover
- Equalisers
- Filters

- High-end audio cables
- Loudspeakers
- Headphones
- Microphones
- Mixers
- Oscillators
- Preamplifiers
- Synthesizers
- Tone controls

Products for the End User

Legacy Products

- phonographic cylinder
- reel-to-reel audio tape
- 8-track tape
- Elcassette

Modern Day Products

Mainstream:

- Turntable for vinyl records (33 1/3 RPM stereo LP)
- Compact disc player
- iPod/MP3/MP4 player

Niche markets:

- Turntables for vinyl records
- Compact cassette
- Minidisc

Analog-to-digital Converter

In electronics, an analog-to-digital converter (ADC, A/D, A–D, or A-to-D) is a system that converts an analog signal, such as a sound picked up by a microphone or light entering a digital camera, into a digital signal. An ADC may also provide an isolated measurement such as an electronic device that converts an input analog voltage or current to a digital number proportional to the magnitude of the voltage or current.

4-channel stereo multiplexed analog-to-digital converter WM8775SEDS made by Wolfson Microelectronics placed on an X-Fi Fatalıty Pro sound card.

Typically the digital output is a two's complement binary number that is proportional to the input, but there are other possibilities.

There are several ADC architectures. Due to the complexity and the need for precisely matched components, all but the most specialized ADCs are implemented as integrated circuits (ICs).

A digital-to-analog converter (DAC) performs the reverse function; it converts a digital signal into an analog signal.

Explanation

The conversion involves quantization of the input, so it necessarily introduces a small amount of error. Furthermore, instead of continuously performing the conversion, an ADC does the conversion periodically, sampling the input. The result is a sequence of digital values that have been converted from a continuous-time and continuous-amplitude analog signal to a discrete-time and discrete-amplitude digital signal.

An ADC is defined by its bandwidth and its signal-to-noise ratio. The bandwidth of an ADC is characterized primarily by its sampling rate. The dynamic range of an ADC is influenced by many factors, including the resolution, linearity and accuracy (how well the quantization levels match the true analog signal), aliasing and jitter. The dynamic range of an ADC is often summarized in terms of its effective number of bits (ENOB), the number of bits of each measure it returns that are on average not noise. An ideal ADC has an ENOB equal to its resolution. ADCs are chosen to match the bandwidth and required signal-to-noise ratio of the signal to be quantized. If an ADC operates at a sampling rate greater than twice the bandwidth of the signal, then perfect reconstruction is possible given an ideal ADC and neglecting quantization error. The presence of quantization error limits the dynamic range of even an ideal ADC. However, if the dynamic range of the ADC exceeds that of the input signal, its effects may be neglected resulting in an essentially perfect digital representation of the input signal.

Resolution

The resolution of the converter indicates the number of discrete values it can produce over the range of analog values. The resolution determines the magnitude of the quantization error and therefore determines the maximum possible average signal to noise ratio for an ideal ADC without the use of oversampling. The values are usually stored electronically in binary form, so the resolution is usually expressed in bits. In consequence, the number of discrete values available, or "levels", is assumed to be a power of two. For example, an ADC with a resolution of 8 bits can encode an analog input to one in 256 different levels, since $2^8 = 256$. The values can represent the ranges from 0 to 255 (i.e. unsigned integer) or from -128 to 127 (i.e. signed integer), depending on the application.

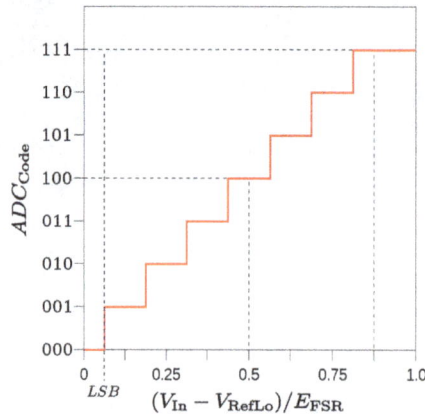

An 8-level ADC coding scheme.

Resolution can also be defined electrically, and expressed in volts. The minimum change in voltage required to guarantee a change in the output code level is called the least significant bit (LSB) voltage. The resolution Q of the ADC is equal to the LSB voltage. The voltage resolution of an ADC is equal to its overall voltage measurement range divided by the number of intervals:

$$Q = \frac{E_{\text{FSR}}}{2^M},$$

where M is the ADC's resolution in bits and E_{FSR} is the full scale voltage range (also called 'span'). E_{FSR} is given by

$$E_{\text{FSR}} = V_{\text{RefHi}} - V_{\text{RefLow}},$$

where V_{RefHi} and V_{RefLow} are the upper and lower extremes, respectively, of the voltages that can be coded.

Normally, the number of voltage intervals is given by

$$N = 2^M,$$

where M is the ADC's resolution in bits.

That is, one voltage interval is assigned in between two consecutive code levels.

Example:

- Coding scheme as in the figure above (assume input signal x(t) = Acos(t), A = 5V)
- Full scale measurement range = -5 to 5 volts

- ADC resolution is 8 bits: 2^8 = 256 quantization levels (codes)

- ADC voltage resolution, Q = (5 V − (-5) V) / 256 = 10 V / 256 ≈ 0.039 V ≈ 39 mV.

In practice, the useful resolution of a converter is limited by the best signal-to-noise ratio (SNR) that can be achieved for a digitized signal. An ADC can resolve a signal to only a certain number of bits of resolution, called the effective number of bits (ENOB). One effective bit of resolution changes the signal-to-noise ratio of the digitized signal by 6 dB, if the resolution is limited by the ADC. If a preamplifier has been used prior to A/D conversion, the noise introduced by the amplifier can be an important contributing factor towards the overall SNR.

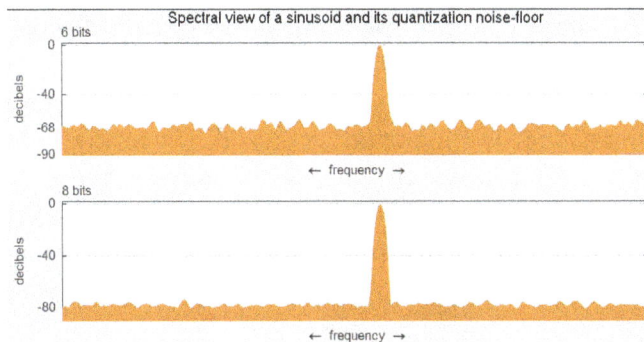

Comparison of quantizing a sinusoid to 64 levels (6 bits) and 256 levels (8 bits). The additive noise created by 6-bit quantization is 12 dB greater than the noise created by 8-bit quantization. When the spectral distribution is flat, as in this example, the 12 dB difference manifests as a measurable difference in the noise floors.

Quantization Error

Quantization error is the noise introduced by quantization in an ideal ADC. It is a rounding error between the analog input voltage to the ADC and the output digitized value. The noise is non-linear and signal-dependent. In an ideal analog-to-digital converter, where the quantization error is uniformly distributed between −1/2 LSB and +1/2 LSB, and the signal has a uniform distribution covering all quantization levels, the Signal-to-quantization-noise ratio (SQNR) can be calculated from

$$\text{SQNR} = 20\log_{10}(2^Q) \approx 6.02 \cdot Q \text{ dB}$$

Where Q is the number of quantization bits. For example, a 16-bit ADC has a maximum signal-to-noise ratio of 6.02 × 16 = 96.3 dB, and therefore the quantization error is 96.3 dB below the maximum level. Quantization error is distributed from DC to the Nyquist frequency, consequently if part of the ADC's bandwidth is not used (as in oversampling), some of the quantization error will fall out of band, effectively improving the SQNR. In an oversampled system, noise shaping can be used to further increase SQNR by forcing more quantization error out of the band.

Dither

In ADCs, performance can usually be improved using dither. This is a very small amount of random noise (white noise), which is added to the input before conversion. Its effect is to cause the state of the LSB to randomly oscillate between 0 and 1 in the presence of very low levels of input, rather than sticking at a fixed value. Rather than the signal simply getting cut off altogether at this low level (which is only being quantized to a resolution of 1 bit), it extends the effective range of signals that the ADC can convert, at the expense of a slight increase in noise – effectively the quantization error is diffused across a series of noise values which is far less objectionable than a hard cutoff. The result is an accurate representation of the signal over time. A suitable filter at the output of the system can thus recover this small signal variation.

An audio signal of very low level (with respect to the bit depth of the ADC) sampled without dither sounds extremely distorted and unpleasant. Without dither the low level may cause the least significant bit to "stick" at 0 or 1. With dithering, the true level of the audio may be calculated by averaging the actual quantized sample with a series of other samples [the dither] that are recorded over time. A virtually identical process, also called dither or dithering, is often used when quantizing photographic images to a fewer number of bits per pixel—the image becomes noisier but to the eye looks far more realistic than the quantized image, which otherwise becomes banded. This analogous process may help to visualize the effect of dither on an analogue audio signal that is converted to digital. Dithering is also used in integrating systems such as electricity meters. Since the values are added together, the dithering produces results that are more exact than the LSB of the analog-to-digital converter.Dither can only increase the resolution of a sampler, it cannot improve the linearity, and thus accuracy does not necessarily improve.

Accuracy

An ADC has several sources of errors. Quantization error and (assuming the ADC is intended to be linear) non-linearity are intrinsic to any analog-to-digital conversion. These errors are measured in a unit called the least significant bit (LSB). In the above example of an eight-bit ADC, an error of one LSB is 1/256 of the full signal range, or about 0.4%.

Non-linearity

All ADCs suffer from non-linearity errors caused by their physical imperfections, causing their output to deviate from a linear function (or some other function, in the case of a deliberately non-linear ADC) of their input. These errors can sometimes be mitigated by calibration, or prevented by testing. Important parameters for linearity are integral non-linearity (INL) and differential non-linearity (DNL). These non-linearities reduce the dynamic range of the signals that can be digitized by the ADC, also reducing the effective resolution of the ADC.

Jitter

When digitizing a sine wave $x(t) = A\sin(2\pi f_0 t)$, the use of a non-ideal sampling clock will result in some uncertainty in when samples are recorded. Provided that the actual sampling time *uncertainty* due to the *clock jitter* is Δt, the error caused by this phenomenon can be estimated as $E_{ap} \leq |x'(t)\Delta t| \leq 2A\pi f_0 \Delta t$. This will result in additional recorded noise that will reduce the effective number of bits (ENOB) below that predicted by quantization error alone. The error is zero for DC, small at low frequencies, but significant when high frequencies have high amplitudes. This effect can be ignored if it is drowned out by the *quantizing error*. Jitter requirements can be calculated using the following formula: $\Delta t < \dfrac{1}{2^q \pi f_0}$, where q is the number of ADC bits.

Output size (bits)	Signal Frequency						
	1 Hz	1 kHz	10 kHz	1 MHz	10 MHz	100 MHz	1 GHz
8	1,243 μs	1.24 μs	124 ns	1.24 ns	124 ps	12.4 ps	1.24 ps
10	311 μs	311 ns	31.1 ns	311 ps	31.1 ps	3.11 ps	0.31 ps
12	77.7 μs	77.7 ns	7.77 ns	77.7 ps	7.77 ps	0.78 ps	0.08 ps
14	19.4 μs	19.4 ns	1.94 ns	19.4 ps	1.94 ps	0.19 ps	0.02 ps
16	4.86 μs	4.86 ns	486 ps	4.86 ps	0.49 ps	0.05 ps	–
18	1.21 μs	1.21 ns	121 ps	1.21 ps	0.12 ps	–	–
20	304 ns	304 ps	30.4 ps	0.30 ps	0.03 ps	–	–

Clock jitter is caused by phase noise. The resolution of ADCs with a digitization bandwidth between 1 MHz and 1 GHz is limited by jitter. When sampling audio signals at 44.1 kHz, the anti-aliasing filter should have eliminated all frequencies above 22 kHz. The input frequency (in this case, < 22 kHz), not the ADC clock frequency, is the determining factor with respect to jitter performance.

Sampling Rate

The analog signal is continuous in time and it is necessary to convert this to a flow of digital values. It is therefore required to define the rate at which new digital values are sampled from the analog signal. The rate of new values is called the *sampling rate* or *sampling frequency* of the converter. A continuously varying bandlimited signal can be sampled (that is, the signal values at intervals of time T, the sampling time, are measured and stored) and then the original signal can be *exactly* reproduced from the discrete-time values by an interpolation formula. The accuracy is limited by quantization error. However, this faithful reproduction is only possible if the sampling rate is higher than twice the highest frequency of the signal. This is essentially what is embodied in the Shannon-Nyquist sampling theorem. Since a practical ADC cannot make an instantaneous conversion, the input value must necessarily be held constant during the time that the converter performs a conversion (called the *conversion time*). An input circuit called a sample and hold performs this task—in most cases by using a capacitor to store the analog voltage at the input, and using an electronic switch or gate to disconnect the capacitor from the input. Many ADC integrated circuits include the sample and hold subsystem internally.

Aliasing

An ADC works by sampling the value of the input at discrete intervals in time. Provided that the input is sampled above the Nyquist rate, defined as twice the highest frequency of interest, then all frequencies in the signal can be reconstructed. If frequencies above half the Nyquist rate are sampled, they are incorrectly detected as lower frequencies, a process referred to as aliasing. Aliasing occurs because instantaneously sampling a function at two or fewer times per cycle results in missed cycles, and therefore the appearance of an incorrectly lower frequency. For example, a 2 kHz sine wave being sampled at 1.5 kHz would be reconstructed as a 500 Hz sine wave.

To avoid aliasing, the input to an ADC must be low-pass filtered to remove frequencies above half the sampling rate. This filter is called an *anti-aliasing filter*, and is essential for a practical ADC system that is applied to analog signals with higher frequency content. In applications where protection against aliasing is essential, oversampling may be used to greatly reduce or even eliminate it. Although aliasing in most systems is unwanted, it should also be noted that it can be exploited to provide simultaneous down-mixing of a band-limited high frequency signal. The alias is effectively the lower heterodyne of the signal frequency and sampling frequency.

Oversampling

Signals are often sampled at the minimum rate required, for economy, with the result that the quantization noise introduced is white noise spread over the whole pass band of the converter. If a signal is sampled at a rate much higher than the Nyquist rate and then digitally filtered to limit it to the signal bandwidth there are the following advantages:

- Digital filters can have better properties (sharper rolloff, phase) than analogue filters, so a sharper anti-aliasing filter can be realised and then the signal can be downsampled giving a better result.

- A 20-bit ADC can be made to act as a 24-bit ADC with 256× oversampling.

- The signal-to-noise ratio due to quantization noise will be higher than if the whole available band had been used. With this technique, it is possible to obtain an effective resolution larger than that provided by the converter alone.

- The improvement in SNR is 3 dB (equivalent to 0.5 bits) per octave of oversampling which is not sufficient for many applications. Therefore, oversampling is usually coupled with noise shaping. With noise shaping, the improvement is 6L+3 dB per octave where L is the order of loop filter used for noise shaping. e.g. – a 2nd order loop filter will provide an improvement of 15 dB/octave.

Oversampling is typically used in audio frequency ADCs where the required sampling rate (typically 44.1 or 48 kHz) is very low compared to the clock speed of typical transistor circuits (>1 MHz). In this case, by using the extra bandwidth to distribute quantization error onto out of band frequencies, the accuracy of the ADC can be greatly increased at no cost. Furthermore, as any aliased signals are also typically out of band, aliasing can often be completely eliminated using very low cost filters.

Relative Speed and Precision

The speed of an ADC varies by type. The Wilkinson ADC is limited by the clock rate which is processable by current digital circuits. Currently, frequencies up to 300 MHz are possible. For a successive-approximation ADC, the conversion time scales with the logarithm of the resolution, e.g. the number of bits. Thus for high resolution, it is possible that the successive-approximation ADC is faster than the Wilkinson. However, the time consuming steps in the Wilkinson are digital, while those in the successive-approximation are analog. Since analog is inherently slower than digital, as the resolution increases, the time required also increases. Thus there are competing processes at work. Flash ADCs are certainly the fastest type of the three. The conversion is basically performed in a single parallel step. For an 8-bit unit, conversion takes place in a few tens of nanoseconds.

There is, as expected, somewhat of a tradeoff between speed and precision. Flash ADCs have drifts and uncertainties associated with the comparator levels. This results in poor linearity. For successive-approximation ADCs, poor linearity is also present, but less so than for flash ADCs. Here, non-linearity arises from accumulating errors from the subtraction processes. Wilkinson ADCs have the highest linearity of the three. These have the best differential non-linearity. The other types require channel smoothing to achieve the level of the Wilkinson.

Sliding Scale Principle

The sliding scale or randomizing method can be employed to greatly improve the linearity of any type of ADC, but especially flash and successive approximation types. For any ADC the mapping from input voltage to digital output value is not exactly a floor or ceiling function as it should be. Under normal conditions, a pulse of a particular amplitude is always converted to a digital value. The problem lies in that the ranges of analog values for the digitized values are not all of the same width, and the differential linearity decreases proportionally with the divergence from the average width. The sliding scale principle uses an averaging effect to overcome this phenomenon. A random, but known analog voltage is added to the sampled input voltage. It is then converted to digital form, and the equivalent digital amount is subtracted, thus restoring it to its original value. The advantage is that the conversion has taken place at a random point. The statistical distribution of the final levels is decided by a weighted average over a region of the range of the ADC. This in turn desensitizes it to the width of any specific level.

Types

These are the most common ways of implementing an electronic ADC:

Direct-conversion

A direct-conversion ADC or flash ADC has a bank of comparators sampling the input signal in parallel, each firing for their decoded voltage range. The comparator bank feeds a logic circuit that generates a code for each voltage range. Direct conversion is very fast, capable of gigahertz sampling rates, but usually has only 8 bits of resolution or fewer, since the number of comparators needed, $2^N - 1$, doubles with each additional bit, requiring a large, expensive circuit. ADCs of this type have a large die size, a high input capacitance, high power dissipation, and are prone to produce glitches at the output (by outputting an out-of-sequence code). Scaling to newer submicrometre technologies does not help as the device mismatch is the dominant design limitation. They are often used for video, wideband communications or other fast signals in optical storage.

Successive Approximation

A successive-approximation ADC uses a comparator to successively narrow a range that contains the input voltage. At each successive step, the converter compares the input voltage to the output of an internal digital to analog converter which might represent the midpoint of a selected voltage range. At each step in this process, the approximation is stored in a successive approximation register (SAR). For example, consider an input voltage of 6.3 V and the initial range is 0 to 16 V. For the first step, the input 6.3 V is compared to 8 V (the midpoint of the

0–16 V range). The comparator reports that the input voltage is less than 8 V, so the SAR is updated to narrow the range to 0–8 V. For the second step, the input voltage is compared to 4 V (midpoint of 0–8). The comparator reports the input voltage is above 4 V, so the SAR is updated to reflect the input voltage is in the range 4–8 V. For the third step, the input voltage is compared with 6 V (halfway between 4 V and 8 V); the comparator reports the input voltage is greater than 6 volts, and search range becomes 6–8 V. The steps are continued until the desired resolution is reached.

Ramp-compare

A ramp-compare ADC produces a saw-tooth signal that ramps up or down then quickly returns to zero. When the ramp starts, a timer starts counting. When the ramp voltage matches the input, a comparator fires, and the timer's value is recorded. Timed ramp converters require the least number of transistors. The ramp time is sensitive to temperature because the circuit generating the ramp is often a simple oscillator. There are two solutions: use a clocked counter driving a DAC and then use the comparator to preserve the counter's value, or calibrate the timed ramp. A special advantage of the ramp-compare system is that comparing a second signal just requires another comparator, and another register to store the voltage value. A very simple (non-linear) ramp-converter can be implemented with a microcontroller and one resistor and capacitor. Vice versa, a filled capacitor can be taken from an integrator, time-to-amplitude converter, phase detector, sample and hold circuit, or peak and hold circuit and discharged. This has the advantage that a slow comparator cannot be disturbed by fast input changes.

Wilkinson

The Wilkinson ADC was designed by D. H. Wilkinson in 1950. The Wilkinson ADC is based on the comparison of an input voltage with that produced by a charging capacitor. The capacitor is allowed to charge until its voltage is equal to the amplitude of the input pulse (a comparator determines when this condition has been reached). Then, the capacitor is allowed to discharge linearly, which produces a ramp voltage. At the point when the capacitor begins to discharge, a gate pulse is initiated. The gate pulse remains on until the capacitor is completely discharged. Thus the duration of the gate pulse is directly proportional to the amplitude of the input pulse. This gate pulse operates a linear gate which receives pulses from a high-frequency oscillator clock. While the gate is open, a discrete number of clock pulses pass through the linear gate and are counted by the address register. The time the linear gate is open is proportional to the amplitude of the input pulse, thus the number of clock pulses recorded in the address register is proportional also. Alternatively, the charging of the capacitor could be monitored, rather than the discharge.

Integrating

An integrating ADC (also dual-slope or multi-slope ADC) applies the unknown input voltage to the input of an integrator and allows the voltage to ramp for a fixed time period (the run-up period). Then a known reference voltage of opposite polarity is applied to the integrator and is allowed to ramp until the integrator output returns to zero (the run-down period). The input voltage is computed as a function of the reference voltage, the constant run-up time period, and the measured run-down time period. The run-down time measurement is usually made in units of the converter's clock, so longer integration times allow for higher resolutions. Likewise, the speed of the converter can be improved by sacrificing resolution. Converters of this type (or variations on the concept) are used in most digital voltmeters for their linearity and flexibility.

Delta-encoded

A delta-encoded ADC or counter-ramp has an up-down counter that feeds a digital to analog converter (DAC). The input signal and the DAC both go to a comparator. The comparator controls the counter. The circuit uses negative feedback from the comparator to adjust the counter until the DAC's output is close enough to the input signal. The number is read from the counter. Delta converters have very wide ranges and high resolution, but the conversion time is dependent on the input signal level, though it will always have a guaranteed worst-case. Delta converters are often very good choices to read real-world signals. Most signals from physical systems do not change abruptly. Some converters combine the delta and successive approximation approaches; this works especially well when high frequencies are known to be small in magnitude.

Pipelined

A pipelined ADC (also called subranging quantizer) uses two or more steps of subranging. First, a coarse conversion is done. In a second step, the difference to the input signal is determined with a digital to analog converter (DAC). This difference is then converted finer, and the results are combined in a last step. This can be considered a refinement of the successive-approximation ADC wherein the feedback reference signal consists of the interim conversion of a whole range of bits (for example, four bits) rather than just the next-most-significant bit. By combining the merits of the successive approximation and flash ADCs this type is fast, has a high resolution, and only requires a small die size.

Sigma-delta

A sigma-delta ADC (also known as a delta-sigma ADC) oversamples the desired signal by a large factor and filters the desired signal band. Generally, a smaller number of bits than required are converted using a Flash ADC after the filter. The resulting signal, along with the error generated by the discrete levels of the Flash, is fed back and subtracted from the input to the filter. This negative feedback has the effect of noise shaping the error due to the Flash so that it does not appear in the desired signal frequencies. A digital filter (decimation filter) follows the ADC which reduces the sampling rate, filters off unwanted noise signal and increases the resolution of the output (sigma-delta modulation, also called delta-sigma modulation).

Time-interleaved

A time-interleaved ADC uses M parallel ADCs where each ADC samples data every M:th cycle of the effective sample clock. The result is that the sample rate is increased M times compared to what each individual ADC can manage. In practice, the individual differences between the M ADCs degrade the overall performance reducing the SFDR. However, technologies exist to correct for these time-interleaving mismatch errors.

Intermediate FM Stage

An ADC with intermediate FM stage first uses a voltage-to-frequency converter to convert the desired signal into an oscillating signal with a frequency proportional to the voltage of the desired signal, and then uses a frequency counter to convert that frequency into a digital count proportional to the desired signal voltage. Longer integration times allow for higher resolutions. Likewise, the speed of the converter can be improved by sacrificing resolution. The two parts of the ADC may be widely separated, with the frequency signal passed through an opto-isolator or transmitted wirelessly. Some such ADCs use sine wave or square wave frequency modulation; others use pulse-frequency modulation. Such ADCs were once the most popular way to show a digital display of the status of a remote analog sensor.

Other Types

There can be other ADCs that use a combination of electronics and other technologies. A time-stretch analog-to-digital converter (TS-ADC) digitizes a very wide bandwidth analog signal, that cannot be digitized by a conventional electronic ADC, by time-stretching the signal prior to digitization. It commonly uses a photonic preprocessor frontend to time-stretch the signal, which effectively slows the signal down in time and compresses its bandwidth. As a result, an electronic backend ADC, that would have been too slow to capture the original signal, can now capture this slowed down signal. For continuous capture of the signal, the frontend also divides the signal into multiple segments in addition to time-stretching. Each segment is individually digitized by a separate electronic ADC. Finally, a digital signal processor rearranges the samples and removes any distortions added by the frontend to yield the binary data that is the digital representation of the original analog signal.

Commercial

Commercial ADCs are usually implemented as integrated circuits. Most converters sample with 6 to 24 bits of resolution, and produce fewer than 1 megasample per second. Thermal noise generated by passive components such as resistors masks the measurement when higher resolution is desired. For audio applications and in room temperatures, such noise is usually a little less than 1 μV (microvolt) of white noise. If the MSB corresponds to a standard 2 V of output signal, this translates to a noise-limited performance that is less than 20~21 bits, and obviates the need for any dithering. As of February 2002, Mega- and giga-sample per second converters are available. Mega-sample converters are required in digital video cameras, video capture cards, and TV tuner cards to convert full-speed analog video to digital video files. Commercial converters usually have ±0.5 to ±1.5 LSB error in their output.

In many cases, the most expensive part of an integrated circuit is the pins, because they make the package larger, and each pin has to be connected to the integrated circuit's silicon. To save pins, it is common for slow ADCs to send their data one bit at a time over a serial interface to the computer, with the next bit coming out when a clock signal changes state, say from 0 to 5 V. This saves quite a few pins on the ADC package, and in many cases, does not make the overall design any more complex (even microprocessors which use memory-mapped I/O only need a few bits of a port to implement a serial bus to an ADC). Commercial ADCs often have several inputs that feed the same converter, usually through an analog multiplexer. Different models of ADC may include sample and hold circuits, instrumentation amplifiers or differential inputs, where the quantity measured is the difference between two voltages.

Applications

Music Recording

Analog-to-digital converters are integral to 2000s era music reproduction technology and digital audio workstation-based sound recording. People often produce music on computers using an analog recording and therefore need analog-to-digital converters to create the pulse-code modulation (PCM) data streams that go onto compact discs and digital music files. The current crop of analog-to-digital converters utilized in music can sample at rates up to 192 kilohertz. Considerable literature exists on these matters, but commercial considerations often play a significant role. Many recording studios record in 24-bit/96 kHz (or higher) pulse-code modulation (PCM) or Direct Stream Digital (DSD) formats, and then downsample or decimate the signal for Red-Book CD production (44.1 kHz) or to 48 kHz for commonly used radio and television broadcast applications.

Digital Signal Processing

People must use ADCs to process, store, or transport virtually any analog signal in digital form. TV tuner cards, for example, use fast video analog-to-digital converters. Slow on-chip 8, 10, 12, or 16 bit analog-to-digital converters are common in microcontrollers. Digital storage oscilloscopes need very fast analog-to-digital converters, also crucial for software defined radio and their new applications.

Scientific Instruments

Digital imaging systems commonly use analog-to-digital converters in digitizing pixels. Some radar systems commonly use analog-to-digital converters to convert signal strength to digital values for subsequent signal processing. Many other in situ and remote sensing systems commonly use analogous technology. The number of binary bits in the resulting digitized numeric values reflects the resolution, the number of unique discrete levels of quantization (signal processing). The correspondence between the analog signal and the digital signal depends on the quantization error. The quantization process must occur at an adequate speed, a constraint that may limit the resolution of the digital signal. Many sensors in scientific instruments produce an analog signal; temperature, pressure, pH, light intensity etc. All these signals can be amplified and fed to an ADC to produce a digital number proportional to the input signal.

Rotary Encoder

Some non-electronic or only partially electronic devices, such as rotary encoders, can also be considered ADCs. Typically the digital output of an ADC will be a two's complement binary number that is proportional to the input. An encoder might output a Gray code.

Electrical Symbol

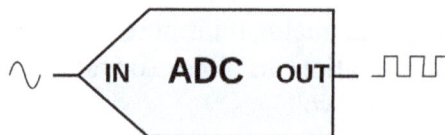

Testing

Testing an Analog to Digital Converter requires an analog input source and hardware to send control signals and capture digital data output. Some ADCs also require an accurate source of reference signal.

The key parameters to test a SAR ADC are:

1. DC offset error

2. DC gain error

3. Signal to noise ratio (SNR)

4. Total harmonic distortion (THD)

5. Integral non linearity (INL)

6. Differential non linearity (DNL)

7. Spurious free dynamic range

8. Power dissipation

Loudspeaker

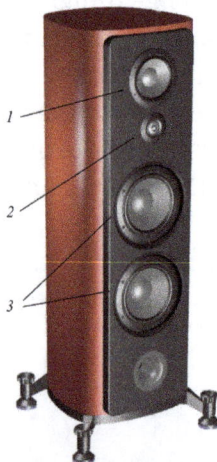

Loudspeaker for home use with three types of dynamic drivers
1. Mid-range driver 2. Tweeter 3. Woofers
The hole below the lowest woofer is a port for a bass reflex system.

A loudspeaker (or loud-speaker or speaker) is an electroacoustic transducer; which converts an electrical audio signal into a corresponding sound. The most widely used type of speaker in the 2010s is the dynamic speaker, invented in 1925 by Edward W. Kellogg and Chester W. Rice. The dynamic speaker operates on the same basic principle as a dynamic microphone, but in reverse, to produce sound from an electrical signal. When an alternating current electrical audio signal is applied to its voice coil, a coil of wire suspended in a circular gap between the poles of a permanent magnet, the coil is forced to move rapidly back and forth due to Faraday's law of induction, which causes a diaphragm (usually conically shaped) attached to the coil to move back and forth, pushing on the air to create sound waves. Besides this most common method, there are several alternative technologies that can be used to convert an electrical signal into sound. The sound source (e.g., a sound recording or a microphone) must be amplified or strengthened with an audio power amplifier before the signal is sent to the speaker.

Speakers are typically housed in a speaker enclosure or speaker cabinet which is often a rectangular or square box made of wood or sometimes plastic. The enclosure's materials and design play an important role in the quality of the sound. Where high fidelity reproduction of sound is required, multiple loudspeaker transducers are often mounted in the same enclosure, each reproducing a part of the audible frequency range *(picture at right)*. In this case the individual speakers are referred to as "drivers" and the entire unit is called a loudspeaker. Drivers made for reproducing high audio frequencies are called tweeters, those for middle frequencies are called mid-range drivers, and those for low frequencies are called woofers. Smaller loudspeakers are found in devices such as radios, televisions, portable audio players, computers, and electronic musical instruments. Larger loudspeaker systems are used for music, sound reinforcement in theatres and concerts, and in public address systems.

Terminology

The term "loudspeaker" may refer to individual transducers (known as "drivers") or to complete speaker systems consisting of an enclosure including one or more drivers.

To adequately reproduce a wide range of frequencies with even coverage, most loudspeaker systems employ more than one driver, particularly for higher sound pressure level or maximum accuracy. Individual drivers are used to reproduce different frequency ranges. The drivers are named subwoofers (for very low frequencies); woofers (low frequencies); mid-range speakers (middle frequencies); tweeters (high frequencies); and sometimes supertweeters, optimized for the highest audible frequencies. The terms for different speaker drivers differ, depending on the application. In two-way systems there is no mid-range driver, so the task of reproducing the mid-range sounds falls upon the woofer and tweeter. Home stereos use the designation "tweeter" for the high frequency driver, while professional concert systems may designate them as "HF" or "highs". When multiple drivers are used in a system, a "filter network", called a crossover, separates the incoming signal into different frequency ranges and routes them to the appropriate driver. A loudspeaker system with n separate frequency bands is described as "n-way speakers": a two-way system will have a woofer and a tweeter; a three-way system employs a woofer, a mid-range, and a tweeter. Loudspeaker driver of the type pictured are termed "dynamic" (short for electrodynamic) to distinguish them from earlier drivers (i.e., moving iron speaker), or speakers using piezoelectric or electrostatic systems, or any of several other sorts.

History

Johann Philipp Reis installed an electric loudspeaker in his *telephone* in 1861; it was capable of reproducing clear tones, but also could reproduce muffled speech after a few revisions. Alexander Graham Bell patented his first electric loudspeaker (capable of reproducing intelligible speech) as part of his telephone in 1876, which was followed in 1877 by an improved version from Ernst Siemens. During this time, Thomas Edison was issued a British patent for a system using compressed air as an amplifying mechanism for his early cylinder phonographs, but he ultimately settled for the familiar metal horn driven by a membrane attached to the stylus. In 1898, Horace Short patented a design for a loudspeaker driven by compressed air; he then sold the rights to Charles Parsons, who was issued several additional British patents before 1910. A few companies, including the Victor Talking Machine Company and Pathé, produced record players using compressed-air loudspeakers. However, these designs were significantly limited by their poor sound quality and their inability to reproduce sound at low volume. Variants of the system were used for public address applications, and more recently, other variations have been used to test space-equipment resistance to the very loud sound and vibration levels that the launching of rockets produces.

The first experimental moving-coil (also called *dynamic*) loudspeaker was invented by Oliver Lodge in 1898. The first practical moving-coil loudspeakers were manufactured by Danish engineer Peter L. Jensen and Edwin Pridham in 1915, in Napa, California. Like previous loudspeakers these used horns to amplify the sound produced by a small diaphragm. Jensen was denied patents. Being unsuccessful in selling their product to telephone companies, in 1915 they changed their target market to radios and public address systems, and named their product Magnavox. Jensen was, for years after the invention of the loudspeaker, a part owner of The Magnavox Company.

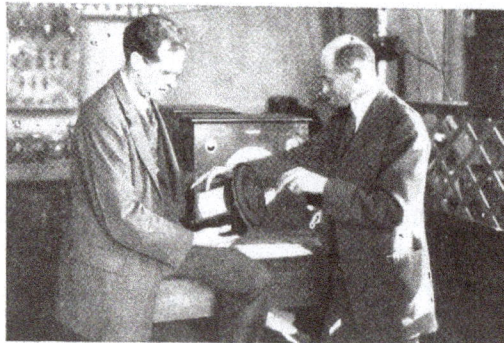

Kellogg and Rice in 1925 holding the large driver of the first moving-coil cone loudspeaker.

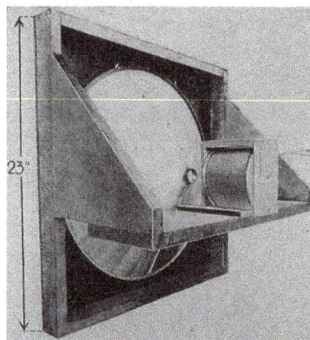

Prototype moving-coil cone loudspeaker by Kellogg and Rice in 1925, with electromagnet pulled back, showing voice coil attached to cone

The moving-coil principle commonly used today in speakers was patented in 1924 by Chester W. Rice and Edward W. Kellogg. The key difference between previous attempts and the patent by Rice and Kellogg is the adjustment of mechanical parameters so that the fundamental resonance of the moving system is below the frequency where the cone's radiation impedance becomes uniform. About this same period, Walter H. Schottky invented the first ribbon loudspeaker together with Dr. Erwin Gerlach.

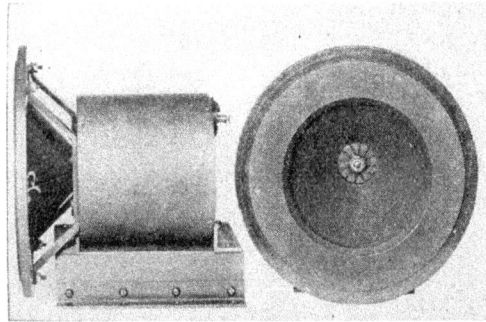

The first commercial version of the speaker, sold with the RCA Radiola receiver, had only a 6 inch cone. In 1926 it sold for $250, equivalent to about $3000 today.

These first loudspeakers used electromagnets, because large, powerful permanent magnets were generally not available at a reasonable price. The coil of an electromagnet, called a field coil, was energized by current through a second pair of connections to the driver. This winding usually served a dual role, acting also as a choke coil, filtering the power supply of the amplifier that the loudspeaker was connected to. AC ripple in the current was attenuated by the action of passing through the choke coil. However, AC line frequencies tended to modulate the audio signal going to the voice coil and added to the audible hum. In 1930 Jensen introduced the first commercial fixed-magnet loudspeaker; however, the large, heavy iron magnets of the day were impractical and field-coil speakers remained predominant until the widespread availability of lightweight Alnico magnets after World War II.

In the 1930s, loudspeaker manufacturers began to combine two and three bandpasses' worth of drivers in order to increase frequency response and sound pressure level. In 1937, the first film industry-standard loudspeaker system, "The Shearer Horn System for Theatres" (a two-way system), was introduced by Metro-Goldwyn-Mayer. It used four 15″ low-frequency drivers, a crossover network set for 375 Hz, and a single multi-cellular horn with two compression drivers providing the high frequencies. John Kenneth Hilliard, James Bullough Lansing, and Douglas Shearer all played roles in creating the system. At the 1939 New York World's Fair, a very large two-way public address system was mounted on a tower at Flushing Meadows. The eight 27″ low-frequency drivers were designed by Rudy Bozak in his role as chief engineer for Cinaudagraph. High-frequency drivers were likely made by Western Electric.

Altec Lansing introduced the *604*, which became their most famous coaxial Duplex driver, in 1943. It incorporated a high-frequency horn that sent sound through the middle of a 15-inch woofer for near-point-source performance. Altec's "Voice of the Theatre" loudspeaker system arrived in the marketplace in 1945, offering better coherence and clarity at the high output levels necessary in movie theaters. The Academy of Motion Picture Arts and Sciences immediately began testing its sonic characteristics; they made it the film house industry standard in 1955.

In 1954, Edgar Villchur developed the acoustic suspension principle of loudspeaker design in Cambridge, Massachusetts. This allowed for better bass response than previously from drivers mounted in smaller cabinets which was important during the transition to stereo recording and reproduction. He and his partner Henry Kloss formed the Acoustic Research company to manufacture and market speaker systems using this principle. Subsequently, continuous developments in enclosure design and materials led to significant audible improvements. The most notable improvements to date in modern dynamic drivers, and the loudspeakers that employ them, are improvements in cone materials, the introduction of higher-temperature adhesives, improved permanent magnet materials, improved measurement techniques, computer-aided design, and finite element analysis. At low frequencies, the application of electrical network theory to the acoustic performance allowed by various enclosure designs (initially by Thiele, and later by Small) has been very important at the design level.

Driver Design: Dynamic Loudspeakers

The most common type of driver, commonly called a dynamic loudspeaker, uses a lightweight diaphragm, or *cone*, connected to a rigid *basket*, or *frame*, via a flexible suspension, commonly called a *spider*, that constrains a voice coil to move axially through a cylindrical magnetic gap. When an electrical signal is applied to the voice coil, a magnetic field is created by the electric current in the voice coil, making it a variable electromagnet. The coil and the driver's magnetic system interact, generating a mechanical force that causes the coil (and thus, the attached cone) to move back and forth, accelerating and reproducing sound under the control of the applied electrical signal coming from the amplifier. The following is a description of the individual components of this type of loudspeaker.

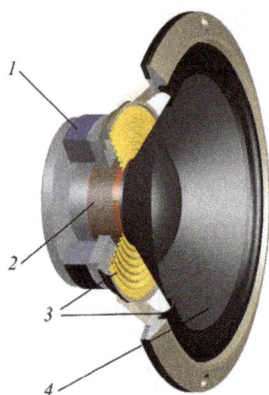

Cutaway view of a dynamic loudspeaker for the bass register.

1. Magnet

2. Voicecoil

3. Suspension

4. Diaphragm

The most common type of driver, commonly called a dynamic loudspeaker, uses a lightweight diaphragm, or *cone*, connected to a rigid *basket*, or *frame*, via a flexible suspension, commonly called a *spider*, that constrains a voice coil to move axially through a cylindrical magnetic gap. When an electrical signal is applied to the voice coil, a magnetic field is created by the electric current in the voice coil, making it a variable electromagnet. The coil and the driver's magnetic system interact, generating a mechanical force that causes the coil (and thus, the attached cone) to move back and forth, accelerating and reproducing sound under the control of the applied electrical signal coming from the amplifier. The following is a description of the individual components of this type of loudspeaker.

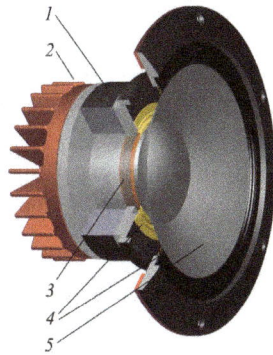

Cutaway view of a dynamic midrange speaker.

1. Magnet

2. Cooler (sometimes present)

3. Voicecoil

4. Suspension 5. Membrane

The diaphragm is usually manufactured with a cone- or dome-shaped profile. A variety of different materials may be used, but the most common are paper, plastic, and metal. The ideal material would 1) be rigid, to prevent uncontrolled cone motions; 2) have low mass, to minimize starting force requirements and energy storage issues; 3) be well damped, to reduce vibrations continuing after the signal has stopped with little or no audible ringing due to its resonance frequency as determined by its usage. In practice, all three of these criteria cannot be met simultaneously using existing materials; thus, driver design involves trade-offs. For example, paper is light and typically well damped, but is not stiff; metal may be stiff and light, but it usually has poor damping; plastic can be light, but typically, the stiffer it is made, the poorer the damping. As a result, many cones are made of some sort of composite material. For example, a cone might be made of cellulose paper, into which some carbon fiber, Kevlar, glass, hemp or bamboo fibers have been added; or it might use a honeycomb sandwich construction; or a coating might be applied to it so as to provide additional stiffening or damping.

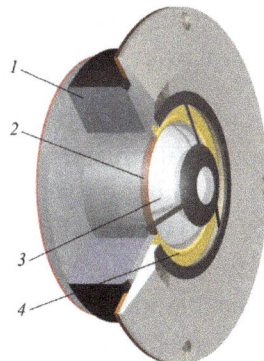

Cutaway view of a dynamic tweeter with acoustic lens and a dome-shaped membrane.

1. Magnet

2. Voicecoil

3. Membrane

4. Suspension

The chassis, frame, or basket, is designed to be rigid, avoiding deformation that could change critical alignments with the magnet gap, perhaps causing the voice coil to rub against the sides of the gap. Chassis are typically cast from aluminum alloy, or stamped from thin steel sheet, though in some drivers with large magnets cast chassis are preferable as sheet metal can easily be warped in whenever the loudspeaker is subjected to rough handling. Other materials such as molded plastic and damped plastic compound baskets are becoming common, especially for inexpensive, low-mass drivers. Metallic chassis can play an important role in conducting heat away from the voice coil; heating during operation changes resistance, causes physical dimensional changes, and if extreme, may even demagnetize permanent magnets.

The suspension system keeps the coil centered in the gap and provides a restoring (centering) force that returns the cone to a neutral position after moving. A typical suspension system consists of two parts: the *spider*, which connects the diaphragm or voice coil to the frame and provides the majority of the restoring force, and the *surround*, which helps center the coil/cone assembly and allows free pistonic motion aligned with the magnetic gap. The spider is usually made of a corrugated fabric disk, impregnated with a stiffening resin. The name comes from the shape of early suspensions, which were two concentric rings of Bakelite material, joined by six or eight curved "legs." Variations of this topology included the addition of a felt disc to provide a barrier to particles that might otherwise cause the voice coil to rub. The German firm Rulik still offers drivers with uncommon spiders made of wood.

The cone surround can be rubber or polyester foam, or a ring of corrugated, resin coated fabric; it is attached to both the outer diaphragm circumference and to the frame. These diverse surround materials, their shape and treatment can dramatically affect the acoustic output of a driver; each implementation having advantages and disadvantages. Polyester foam, for example, is lightweight and economical, though usually leaking air to some degree, but is degraded by exposure to ozone, UV light, humidity and elevated temperatures, significantly limiting useful life with adequate performance.

The wire in a voice coil is usually made of copper, though aluminum—and, rarely, silver—may be used. The advantage of aluminum is its light weight, which reduces the moving mass compared to copper. This raises the resonant frequency of the speaker and increases its efficiency. A disadvantage of aluminum is that it is not easily soldered, and so connections are instead often crimped together and sealed. These connections must be made well or they may fail in an intense environment of mechanical vibration. Voice-coil wire cross sections can be circular, rectangular, or hexagonal, giving varying amounts of wire volume coverage in the magnetic gap space. The coil is oriented co-axially inside the gap; it moves back and forth within a small circular volume (a hole, slot, or groove) in the magnetic structure. The gap establishes a concentrated magnetic field between the two poles of a permanent magnet; the outside of the gap being one pole, and the center post (called the pole piece) being the other. The pole piece and backplate are often a single piece, called the poleplate or yoke.

Modern driver magnets are almost always permanent and made of ceramic, ferrite, Alnico, or, more recently, rare earth such as neodymium and samarium cobalt. Electrodynamic drivers were often used in musical instrument amplifier/speaker cabinets well into the 1950s; there were economic savings in those using tube amplifiers as the field coil could, and usually did, do double duty as a power supply choke. A trend in design — due to increases in transportation costs and a desire for smaller, lighter devices (as in many home theater multi-speaker installations) — is the use of the last instead of heavier ferrite types. Very few manufacturers still produce electrodynamic loudspeakers with electrically powered field coils, as was common in the earliest designs; one of the last is a French firm. When high field-strength permanent magnets became available after WWII, Alnico, an alloy of aluminum, nickel, and cobalt became popular, since it dispensed with the problems of field-coil drivers. Alnico was used almost exclusively until about 1980, despite the embarrassing problem of Alnico magnets being partially degaussed (i.e., demagnetized) by accidental 'pops' or 'clicks' caused by loose connections, especially if used with a high-power amplifier. The damage can be reversed by "recharging" the magnet, but this requires uncommon specialist equipment and knowledge.

After 1980, most (but not quite all) driver manufacturers switched from Alnico to ferrite magnets, which are made from a mix of ceramic clay and fine particles of barium or strontium ferrite. Although the energy per kilogram of these ceramic magnets is lower than Alnico, it is substantially less expensive, allowing designers to use larger yet more economical magnets to achieve a given performance.

The size and type of magnet and details of the magnetic circuit differ, depending on design goals. For instance, the shape of the pole piece affects the magnetic interaction between the voice coil and the magnetic field, and is sometimes used to modify a driver's behavior. A "shorting ring", or Faraday loop, may be included as a thin copper cap fitted over the pole tip or as a heavy ring situated within the magnet-pole cavity. The benefits of this complication is reduced impedance at high frequencies, providing extended treble output, reduced harmonic distortion, and a reduction in the inductance modulation that typically accompanies large voice coil excursions. On the other hand, the copper cap requires a wider voice-coil gap, with increased magnetic reluctance; this reduces available flux, requiring a larger magnet for equivalent performance.

Driver design—including the particular way two or more drivers are combined in an enclosure to make a speaker system—is both an art, involving subjective perceptions of timbre and sound quality and a science, involving measurements and experiments. Adjusting a design to improve performance is done using a combination of magnetic, acoustic, mechanical, electrical, and material science theory, and tracked with high precision measurements and the observations of experienced listeners. A few of the issues speaker and driver designers must confront are distortion, radiation lobing, phase effects, off-axis response, and crossover artifacts. Designers can use an anechoic chamber to ensure the speaker can be measured independently of room effects, or any of several electronic techniques that, to some extent, substitute for such chambers. Some developers eschew anechoic chambers in favor of specific standardized room setups intended to simulate real-life listening conditions.

Fabrication of finished loudspeaker systems has become segmented, depending largely on price, shipping costs, and weight limitations. High-end speaker systems, which are typically heavier (and often larger) than economic shipping allows outside local regions, are usually made in their target market region and can cost $140,000 or more per pair. Economical mass market speaker systems and drivers available for much lower costs may be manufactured in China or other low-cost manufacturing locations.

Driver Types

A four-way, high fidelity loudspeaker system. Each of the four drivers outputs a different frequency range; the fifth aperture at the bottom is a bass reflex port

Individual electrodynamic drivers provide their best performance within a limited frequency range. Multiple drivers (e.g., subwoofers, woofers, mid-range drivers, and tweeters) are generally combined into a complete loudspeaker system to provide performance beyond that constraint. The three most commonly used sound radiation systems are the cone, dome and horn type drivers.

Full-range Drivers

A full-range driver is a speaker designed to be used alone to reproduce an audio channel without the help of other drivers, and therefore must cover the entire audio frequency range. These drivers are small, typically 3 to 8 inches (7.6 to 20.3 cm) in diameter to permit reasonable high frequency response, and carefully designed to give low-distortion output at low frequencies, though with reduced maximum output level. Full-range (or more accurately, wide-range) drivers are most commonly heard in public address systems, in televisions (although some models are suitable for hi-fi listening), small radios, intercoms, some computer speakers, etc. In hi-fi speaker systems, the use of wide-range drive units can avoid undesirable interactions between multiple drivers caused by non-coincident driver location or crossover network issues. Fans of wide-range driver hi-fi speaker systems claim a coherence of sound due to the single source and a resulting lack of interference, and likely also to the lack of crossover components. Detractors typically cite wide-range drivers' limited frequency response and modest output abilities (most especially at low frequencies), together with their requirement for large, elaborate, expensive enclosures—such as transmission lines, quarter wave resonators or horns—to approach optimum performance. With the advent of neodymium drivers, low cost quarter wave transmission lines are made possible and are increasingly made availably commercially.

Full-range drivers often employ an additional cone called a *whizzer*: a small, light cone attached to the joint between the voice coil and the primary cone. The whizzer cone extends the high-frequency response of the driver and broadens its high frequency directivity, which would otherwise be greatly narrowed due to the outer diameter cone material failing to keep up with the central voice coil at higher frequencies. The main cone in a whizzer design is manufactured so as to flex more in the outer diameter than in the center. The result is that the main cone delivers low frequencies and the whizzer cone contributes most of the higher frequencies. Since the whizzer cone is smaller than the main diaphragm, output dispersion at high frequencies is improved relative to an equivalent single larger diaphragm.

Limited-range drivers, also used alone, are typically found in computers, toys, and clock radios. These drivers are less elaborate and less expensive than wide-range drivers, and they may be severely compromised to fit into very small mounting locations. In these applications, sound quality is a low priority. The human ear is remarkably tolerant of poor sound quality, and the distortion inherent in limited-range drivers may enhance their output at high frequencies, increasing clarity when listening to spoken word material.

Subwoofer

A subwoofer is a woofer driver used only for the lowest-pitched part of the audio spectrum: typically below 200 Hz for consumer systems, below 100 Hz for professional live sound, and below 80 Hz in THX-approved systems. Because the intended range of frequencies is limited, subwoofer system design is usually simpler in many respects than for conventional loudspeakers, often consisting of a single driver enclosed in a suitable box or enclosure. Since sound in this frequency range can easily bend around corners by diffraction, the speaker aperture does not have to face the audience, and subwoofers can be mounted in the bottom of the enclosure, facing the floor. This is eased by the limitations of human hearing at low frequencies; such sounds cannot be located in space, due to their large wavelengths compared to higher frequencies which produce differential effects in the ears due to shadowing by the head, and diffraction around it, both of which we rely upon for localization clues.

To accurately reproduce very low bass notes without unwanted resonances (typically from cabinet panels), subwoofer systems must be solidly constructed and properly braced to avoid unwanted sounds of cabinet vibrations. As a result, good subwoofers are typically quite heavy. Many subwoofer systems include integrated power amplifiers and electronic subsonic (sub)-filters, with additional controls relevant to low-frequency reproduction (e.g., a crossover knob and a phase switch). These variants are known as "active" or "powered" subwoofers, with the former including a power amplifier. In contrast, "passive" subwoofers require external amplification.

In typical installations, subwoofers are physically separated from the rest of the speaker cabinets. Because of propagation delay, their output may be somewhat out of phase from another subwoofer (on another channel) or slightly out of phase with the rest of the sound. Consequently, a subwoofer's power amp often has a phase-delay adjustment (approximately 1 ms of delay is required for each additional foot of separation from the listener) which may improve performance of the system as a whole at subwoofer frequencies (and perhaps an octave or so above the crossover point). However, the influence of room resonances (sometimes called standing waves) is typically so large that such issues are secondary in practice. Subwoofers are widely used in large concert and mid-sized venue sound reinforcement systems. Subwoofer cabinets are often built with a bass reflex port (i.e., a hole cut into the cabinet with a tube attached to it), a design feature which if properly engineered improves bass performance and increases efficiency.

Woofer

A woofer is a driver that reproduces low frequencies. The driver works with the characteristics of the enclosure to produce suitable low frequencies. Indeed, both are so closely connected that they must be considered together in use. Only at design time do the separate properties of enclosure and woofer matter individually. Some loudspeaker systems use a woofer for the lowest frequencies, sometimes well enough that a subwoofer is not needed. Additionally, some loudspeakers use the woofer to handle middle frequencies, eliminating the mid-range driver. This can be accomplished with the selection of a tweeter that can work low enough that, combined with a woofer that responds high enough, the two drivers add coherently in the middle frequencies.

Mid-range Driver

A mid-range speaker is a loudspeaker driver that reproduces a band of frequencies generally between 1–6 kHz, otherwise known as the 'mid' frequencies (between the woofer and tweeter). Mid-range driver diaphragms can be made of paper or composite materials, and can be direct radiation drivers (rather like smaller woofers) or they can be compression drivers (rather like some tweeter designs). If the mid-range driver is a direct radiator, it can be mounted on the front baffle of a loudspeaker enclosure, or, if a compression driver, mounted at the throat of a horn for added output level and control of radiation pattern.

Tweeter

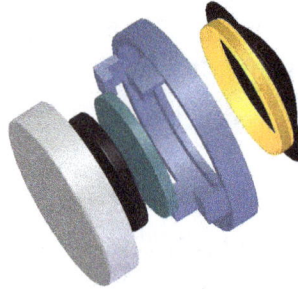

Exploded view of a dome tweeter.

A tweeter is a high-frequency driver that reproduces the highest frequencies in a speaker system. A major problem in tweeter design is achieving wide angular sound coverage (off-axis response), since high frequency sound tends to leave the speaker in narrow beams. Soft-dome tweeters are widely found in home stereo systems, and horn-loaded compression drivers are common in professional sound reinforcement. Ribbon tweeters have gained popularity in recent years, as the output power of some designs has been increased to levels useful for professional sound reinforcement, and their output pattern is wide in the horizontal plane, a pattern that has convenient applications in concert sound.

Coaxial Drivers

A coaxial driver is a loudspeaker driver with two or several combined concentric drivers. Coaxial drivers have been produced by many companies, such as Altec, Tannoy, Pioneer, KEF, SEAS, B&C Speakers, BMS, Cabasse and Genelec.

System Design

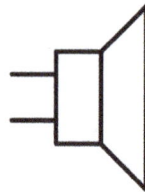

Electronic symbol for a speaker

Crossover

Used in multi-driver speaker systems, the crossover is an assembly of filters that separate the input signal into different frequency ranges (i.e. "bands"), according to the requirements of each driver. Hence the drivers receive power only at their operating frequency (the sound frequency range they were designed for), thereby reducing distortion in the drivers and interference between them. The ideal characteristics of a crossover may include perfect out-of-band attenuation at the output of each filter, no amplitude variation ("ripple") within each passband, no phase delay between overlapping frequency bands, to name just a few.

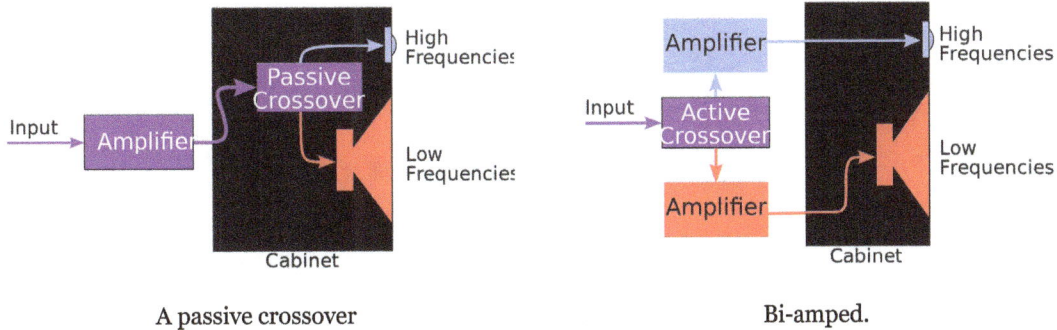

A passive crossover   Bi-amped.

Crossovers can be *passive* or *active*. A passive crossover is an electronic circuit that uses a combination of one or more resistors, inductors, or non-polar capacitors. These components are combined to form a filter network and are most often placed between the full frequency-range power amplifier and the loudspeaker drivers to divide the amplifier's signal into the necessary frequency bands before being delivered to the individual drivers. Passive crossover circuits need no external power beyond the audio signal itself, but have some disadvantages: they may require larger inductors and capacitors due to power handling requirements (being driven by the amplifier), limited component availability to optimize the crossover's characteristics at such power levels, etc. Unlike active crossovers which include a built-in amplifier, passive crossovers have an inherent attenuation within the passband, typically leading to a reduction in damping factor before the voice coil An active crossover is an electronic filter circuit that divides the signal into individual frequency bands *before* power amplification, thus requiring at least one power amplifier for each bandpass. Passive filtering may also be used in this way before power amplification, but it is an uncommon solution, being less flexible than active filtering. Any technique that uses crossover filtering followed by amplification is commonly known as bi-amping, tri-amping, quad-amping, and so on, depending on the minimum number of amplifier channels.

Some loudspeaker designs use a combination of passive and active crossover filtering, such as a passive crossover between the mid- and high-frequency drivers and an active crossover between the low-frequency driver and the combined mid- and high frequencies.

Passive crossovers are commonly installed inside speaker boxes and are by far the most usual type of crossover for home and low-power use. In car audio systems, passive crossovers may be in a separate box, necessary to accommodate the size of the components used. Passive crossovers may be simple for low-order filtering, or complex to allow steep slopes such as 18 or 24 dB per octave. Passive crossovers can also be designed to compensate for undesired characteristics of driver, horn, or enclosure resonances, and can be tricky to implement, due to component interaction. Passive crossovers, like the driver units that they feed, have power handling limits, have insertion losses (10% is often claimed), and change the load seen by the amplifier. The changes are matters of concern for many in the hi-fi world. When high output levels are required, active crossovers may be preferable. Active crossovers may be simple circuits that emulate the response of a passive network, or may be more complex, allowing extensive audio adjustments. Some active crossovers, usually digital loudspeaker management systems, may include electronics and controls for precise alignment of phase and time between frequency bands, equalization, dynamic range compression and limiting) control.

Enclosures

An unusual three-way speaker system. The cabinet is narrow to raise the frequency
where a diffraction effect called the "baffle step" occurs.

Most loudspeaker systems consist of drivers mounted in an enclosure, or cabinet. The role of the
enclosure is to prevent sound waves emanating from the back of a driver from interfering destruc-
tively with those from the front. The sound waves emitted from the back are 180° out of phase with
those emitted forward, so without an enclosure they typically cause cancellations which significantly degrade the level and quality of sound at low frequencies.

The simplest driver mount is a flat panel (i.e., baffle) with the drivers mounted in holes in it. However, in this approach, sound frequencies with a wavelength longer than the baffle dimensions are
canceled out, because the antiphase radiation from the rear of the cone interferes with the radiation from the front. With an infinitely large panel, this interference could be entirely prevented. A
sufficiently large sealed box can approach this behavior.

Since panels of infinite dimensions are impossible, most enclosures function by containing the rear
radiation from the moving diaphragm. A sealed enclosure prevents transmission of the sound emitted from the rear of the loudspeaker by confining the sound in a rigid and airtight box. Techniques
used to reduce transmission of sound through the walls of the cabinet include thicker cabinet walls,
lossy wall material, internal bracing, curved cabinet walls—or more rarely, visco-elastic materials
(e.g., mineral-loaded bitumen) or thin lead sheeting applied to the interior enclosure walls.

However, a rigid enclosure reflects sound internally, which can then be transmitted back through
the loudspeaker diaphragm—again resulting in degradation of sound quality. This can be reduced
by internal absorption using absorptive materials (often called "damping"), such as glass wool,
wool, or synthetic fiber batting, within the enclosure. The internal shape of the enclosure can also
be designed to reduce this by reflecting sounds away from the loudspeaker diaphragm, where they
may then be absorbed.

Other enclosure types alter the rear sound radiation so it can add constructively to the output from the front of the cone. Designs that do this (including *bass reflex*, *passive radiator*, *transmission line*, etc.) are often used to extend the effective low-frequency response and increase low-frequency output of the driver.

To make the transition between drivers as seamless as possible, system designers have attempted to time-align (or phase adjust) the drivers by moving one or more driver mounting locations forward or back so that the acoustic center of each driver is in the same vertical plane. This may also involve tilting the face speaker back, providing a separate enclosure mounting for each driver, or (less commonly) using electronic techniques to achieve the same effect. These attempts have resulted in some unusual cabinet designs.

The speaker mounting scheme (including cabinets) can also cause diffraction, resulting in peaks and dips in the frequency response. The problem is usually greatest at higher frequencies, where wavelengths are similar to, or smaller than, cabinet dimensions. The effect can be minimized by rounding the front edges of the cabinet, curving the cabinet itself, using a smaller or narrower enclosure, choosing a strategic driver arrangement, using absorptive material around a driver, or some combination of these and other schemes.

Horn Loudspeakers

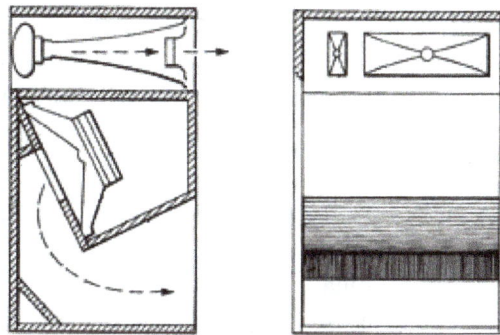

A three-way loudspeaker that uses horns in front of each of the three drivers:
a shallow horn for the tweeter, a long, straight horn for mid frequencies and a folded horn for the woofer

Horn loudspeakers are the oldest form of loudspeaker system. The use of horns as voice-amplifying megaphones dates at least to the 17th century, and horns were used in mechanical gramophones as early as 1857. Horn loudspeakers use a shaped waveguide in front of or behind the driver to increase the directivity of the loudspeaker and to transform a small diameter, high pressure condition at the driver cone surface to a large diameter, low pressure condition at the mouth of the horn. This improves the acoustic -- electro/mechanical impedance match between the driver and ambient air, increasing efficiency, and focusing the sound over a narrower area.

The size of the throat, mouth, the length of the horn, as well as the area expansion rate along it must be carefully chosen to match the drive to properly provide this transforming function over a range of frequencies (every horn performs poorly outside its acoustic limits, at both high and low frequencies). The length and cross-sectional mouth area required to create a bass or sub-bass horn require a horn many feet long. 'Folded' horns can reduce the total size, but compel designers to make compromises and accept increased complication such as cost and construction. Some horn designs not only fold the low frequency horn, but use the walls in a room corner as an extension of the horn mouth. In the late 1940s, horns whose mouths took up much of a room wall were not unknown amongst hi-fi fans. Room sized installations became much less acceptable when two or more were required.

A horn loaded speaker can have a sensitivity as high as 110 dB at 2.83 volts (1 watt at 8 ohms) at 1 meter. This is a hundredfold increase in output compared to a speaker rated at 90 dB sensitivity, and is invaluable in applications where high sound levels are required or amplifier power is limited.

Wiring Connections

Two-way binding posts on a loudspeaker, connected using banana plugs.

Most home hi-fi loudspeakers use two wiring points to connect to the source of the signal (for example, to the audio amplifier or receiver). To accept the wire connection, the loudspeaker enclosure may have binding posts, spring clips, or a panel-mount jack. If the wires for a pair of speakers are not connected with respect to the proper electrical polarity (the + and − connections on the speaker and amplifier should be connected + to + and − to −; speaker cable is almost always marked so that one conductor of a pair can be distinguished from the other, even if it has run under or behind things in its run from amplifier to speaker location), the loudspeakers are said to be "out of phase" or more properly "out of polarity". Given identical signals, motion in one cone is in the opposite direction of the other. This typically causes monophonic material in a stereo recording to be canceled out, reduced in level, and made more difficult to localize, all due to destructive interference of the sound waves. The cancellation effect is most noticeable at frequencies where the loudspeakers are separated by a quarter wavelength or less; low frequencies are affected the most. This type of miswiring error does not damage speakers, but is not optimal for listening.

A 4-ohm loudspeaker with two pairs of binding posts capable of accepting bi-wiring
after the removal of two metal straps.

With sound reinforcement system, PA system and instrument amplifier speaker enclosures, cables and some type of jack or connector are typically used. Lower- and mid-priced sound system and instrument speaker cabinets often use 1/4" speaker cable jacks. Higher-priced and higher powered sound system cabinets and instrument speaker cabinets often use Speakon connectors. Speakon connectors are considered to be safer for high wattage amplifiers, because the connector is designed so that human users cannot touch the connectors.

Wireless Speakers

HP Roar Wireless Speaker

Wireless speakers are very similar to traditional (wired) loudspeakers, but they receive audio signals using radio frequency (RF) waves rather than over audio cables. There is normally an amplifier integrated in the speaker's cabinet because the RF waves alone are not enough to drive the speaker. This integration of amplifier and loudspeaker is known as an active loudspeaker. Manufacturers of these loudspeakers design them to be as lightweight as possible while producing the maximum amount of audio output efficiency.

Wireless speakers still need power, so require a nearby AC power outlet, or possibly batteries. Only the wire to the amplifier is eliminated.

Specifications

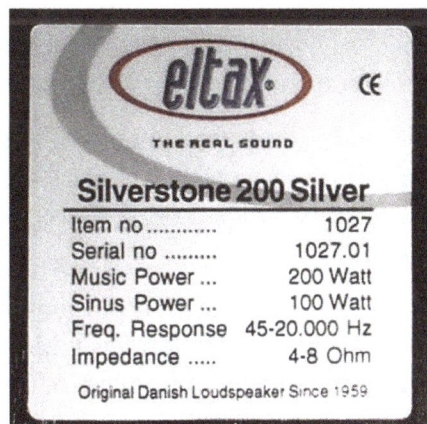

Specifications label on a loudspeaker.

Speaker specifications generally include:

- Speaker or driver type (individual units only) – Full-range, woofer, tweeter, or mid-range.

- Size of individual drivers. For cone drivers, the quoted size is generally the outside diameter of the basket. However, it may less commonly also be the diameter of the cone surround, measured apex to apex, or the distance from the center of one mounting hole to its opposite. Voice-coil diameter may also be specified. If the loudspeaker has a compression horn driver, the diameter of the horn throat may be given.

- Rated Power – Nominal (or even continuous) power, and peak (or maximum short-term) power a loudspeaker can handle (i.e., maximum input power before destroying the loudspeaker; it is never the sound output the loudspeaker produces). A driver may be damaged at much less than its rated power if driven past its mechanical limits at lower frequencies. Tweeters can also be damaged by amplifier clipping (amplifier circuits produce large amounts of energy at high frequencies in such cases) or by music or sine wave input at high frequencies. Each of these situations might pass more energy to a tweeter than it can survive without damage. In some jurisdictions, power handling has a legal meaning allowing comparisons between loudspeakers under consideration. Elsewhere, the variety of meanings for power handling capacity can be quite confusing.

- Impedance – typically 4 Ω (ohms), 8 Ω, etc.

- Baffle or enclosure type (enclosed systems only) – Sealed, bass reflex, etc.

- Number of drivers (complete speaker systems only) – two-way, three-way, etc.

and optionally:

- Crossover frequency(ies) (multi-driver systems only) – The nominal frequency boundaries of the division between drivers.

- Frequency response – The measured, or specified, output over a specified range of frequencies for a constant input level varied across those frequencies. It sometimes includes a variance limit, such as within "± 2.5 dB."

- Thiele/Small parameters (individual drivers only) – these include the driver's F_s (resonance frequency), Q_{ts} (a driver's Q; more or less, its damping factor at resonant frequency), V_{as} (the equivalent air compliance volume of the driver), etc.

- Sensitivity – The sound pressure level produced by a loudspeaker in a non-reverberant environment, often specified in dB and measured at 1 meter with an input of 1 watt (2.83 rms volts into 8 Ω), typically at one or more specified frequencies. Manufacturers often use this rating in marketing material.

- Maximum sound pressure level – The highest output the loudspeaker can manage, short of damage or not exceeding a particular distortion level. Manufacturers often use this rating in marketing material—commonly without reference to frequency range or distortion level.

Electrical Characteristics of dynamic Loudspeakers

The load that a driver presents to an amplifier consists of a complex electrical impedance—a combination of resistance and both capacitive and inductive reactance, which combines properties of the driver, its mechanical motion, the effects of crossover components (if any are in the signal path between amplifier and driver), and the effects of air loading on the driver as modified by the enclosure and its environment. Most amplifiers' output specifications are given at a specific power into an ideal resistive load; however, a loudspeaker does not have a constant resistance across its frequency range. Instead, the voice coil is inductive, the driver has mechanical resonances, the enclosure changes the driver's electrical and mechanical characteristics, and a passive crossover between the drivers and the amplifier contributes its own variations. The result is a load resistance that varies widely with frequency, and usually a varying phase relationship between voltage and current as well, also changing with frequency. Some amplifiers can cope with the variation better than others can.

To make sound, a loudspeaker is driven by modulated electric current (produced by an amplifier) that pass through a "speaker coil" which then (through inductance) creates a magnetic field around the coil, creating a magnetic field. The electric current variations that pass through the speaker are thus converted to a varying magnetic field, whose interaction with the driver's magnetic field moves the speaker diaphragm, which thus forces the driver to produce air motion that is similar to the original signal from the amplifier.

Electromechanical Measurements

Examples of typical measurements are: amplitude and phase characteristics vs. frequency; impulse response under one or more conditions (e.g., square waves, sine wave bursts, etc.); directivity vs. frequency (e.g., horizontally, vertically, spherically, etc.); harmonic and intermodulation distortion vs. sound pressure level (SPL) output, using any of several test signals; stored energy (i.e., ringing) at various frequencies; impedance vs. frequency; and small-signal vs. large-signal performance. Most of these measurements require sophisticated and often expensive equipment to perform, and also good judgment by the operator, but the raw sound pressure level output is rather easier to report and so is often the only specified value—sometimes in misleadingly exact terms. The sound pressure level (SPL) a loudspeaker produces is measured in decibels (dB_{spl}).

Efficiency vs. Sensitivity

Loudspeaker efficiency is defined as the sound power output divided by the electrical power input. Most loudspeakers are inefficient transducers; only about 1% of the electrical energy sent by an amplifier to a typical home loudspeaker is converted to acoustic energy. The remainder is converted to heat, mostly in the voice coil and magnet assembly. The main reason for this is the difficulty of achieving proper impedance matching between the acoustic impedance of the drive unit and the air it radiates into. (At low frequencies, improving this match is the main purpose of speaker enclosure designs). The efficiency of loudspeaker drivers varies with frequency as well. For instance, the output of a woofer driver decreases as the input frequency decreases because of the increasingly poor match between air and the driver.

Driver ratings based on the SPL for a given input are called sensitivity ratings and are notionally similar to efficiency. Sensitivity is usually defined as so many decibels at 1 W electrical input, measured at 1 meter (except for headphones), often at a single frequency. The voltage used is often 2.83 V_{RMS}, which is 1 watt into an 8 Ω (nominal) speaker impedance (approximately true for many speaker systems). Measurements taken with this reference are quoted as dB with 2.83 V @ 1 m.

The sound pressure output is measured at (or mathematically scaled to be equivalent to a measurement taken at) one meter from the loudspeaker and on-axis (directly in front of it), under the condition that the loudspeaker is radiating into an infinitely large space and mounted on an infinite baffle. Clearly then, sensitivity does not correlate precisely with efficiency, as it also depends on the directivity of the driver being tested and the acoustic environment in front of the actual loudspeaker. For example, a cheerleader's horn produces more sound output in the direction it is pointed by concentrating sound waves from the cheerleader in one direction, thus "focusing" them. The horn also improves impedance matching between the voice and the air, which produces more acoustic power for a given speaker power. In some cases, improved impedance matching (via careful enclosure design) lets the speaker produce more acoustic power.

- Typical home loudspeakers have sensitivities of about 85 to 95 dB for 1 W @ 1 m—an efficiency of 0.5–4%.

- Sound reinforcement and public address loudspeakers have sensitivities of perhaps 95 to 102 dB for 1 W @ 1 m—an efficiency of 4–10%.

- Rock concert, stadium PA, marine hailing, etc. speakers generally have higher sensitivities of 103 to 110 dB for 1 W @ 1 m—an efficiency of 10–20%.

A driver with a higher maximum power rating cannot necessarily be driven to louder levels than a lower-rated one, since sensitivity and power handling are largely independent properties. In the examples that follow, assume (for simplicity) that the drivers being compared have the same electrical impedance, are operated at the same frequency within both driver's respective pass bands, and that power compression and distortion are low. For the first example, a speaker 3 dB more sensitive than another produces double the sound power (is 3 dB louder) for the same power input. Thus, a 100 W driver ("A") rated at 92 dB for 1 W @ 1 m sensitivity puts out twice as much acoustic power as a 200 W driver ("B") rated at 89 dB for 1 W @ 1 m when both are driven with 100 W of input power. In this particular example, when driven at 100 W, speaker A produces the same SPL, or loudness as speaker B would produce with 200 W input. Thus, a 3 dB increase in sensitivity of the speaker means that it needs half the amplifier power to achieve a given SPL. This translates into a smaller, less complex power amplifier—and often, to reduced overall system cost.

It is typically not possible to combine high efficiency (especially at low frequencies) with compact enclosure size and adequate low frequency response. One can, for the most part, choose only two of the three parameters when designing a speaker system. So, for example, if extended low-frequency performance and small box size are important, one must accept low efficiency. This rule of thumb is sometimes called Hofmann's Iron Law (after J.A. Hofmann, the "H" in KLH).

Listening Environment

The interaction of a loudspeaker system with its environment is complex and is largely out of the loudspeaker designer's control. Most listening rooms present a more or less reflective environment, depending on size, shape, volume, and furnishings. This means the sound reaching a listener's ears consists not only of sound directly from the speaker system, but also the same sound delayed by traveling to and from (and being modified by) one or more surfaces. These reflected sound waves, when added to the direct sound, cause cancellation and addition at assorted frequencies (e.g., from resonant room modes), thus changing the timbre and character of the sound at the listener's ears. The human brain is very sensitive to small variations, including some of these, and this is part of the reason why a loudspeaker system sounds different at different listening positions or in different rooms.

At Jay Pritzker Pavilion, a LARES system is combined with a zoned sound reinforcement system, both suspended on an overhead steel trellis, to synthesize an indoor acoustic environment outdoors.

A significant factor in the sound of a loudspeaker system is the amount of absorption and diffusion present in the environment. Clapping one's hands in a typical empty room, without draperies or carpet, produces a zippy, fluttery echo due both to a lack of absorption and to reverberation (that is, repeated echoes) from flat reflective walls, floor, and ceiling. The addition of hard surfaced furniture, wall hangings, shelving and even baroque plaster ceiling decoration changes the echoes, primarily because of diffusion caused by reflective objects with shapes and surfaces having sizes on the order of the sound wavelengths. This somewhat breaks up the simple reflections otherwise caused by bare flat surfaces, and spreads the reflected energy of an incident wave over a larger angle on reflection.

Placement

In a typical rectangular listening room, the hard, parallel surfaces of the walls, floor and ceiling cause primary acoustic resonance nodes in each of the three dimensions: left-right, up-down and forward-backward. Furthermore, there are more complex resonance modes involving three, four, five and even all six boundary surfaces combining to create standing waves. Low frequencies excite these modes the most, since long wavelengths are not much affected by furniture compositions or placement. The mode spacing is critical, especially in small and medium size rooms like recording studios, home theaters and broadcast studios. The proximity of the loudspeakers to room boundaries affects how strongly the resonances are excited as well as affecting the relative strength at each frequency. The location of the listener is critical, too, as a position near a boundary can have a great effect on the perceived balance of frequencies. This is because standing wave patterns are most easily heard in these locations and at lower frequencies, below the Schroeder frequency – typically around 200–300 Hz, depending on room size.

Directivity

Acousticians, in studying the radiation of sound sources have developed some concepts important to understanding how loudspeakers are perceived. The simplest possible radiating source is a point source, sometimes called a simple source. An ideal point source is an infinitesimally small point radiating sound. It may be easier to imagine a tiny pulsating sphere, uniformly increasing and decreasing in diameter, sending out sound waves in all directions equally, independent of frequency.

Any object radiating sound, including a loudspeaker system, can be thought of as being composed of combinations of such simple point sources. The radiation pattern of a combination of point sources is not the same as for a single source, but depends on the distance and orientation between the sources, the position relative to them from which the listener hears the combination, and the frequency of the sound involved. Using geometry and calculus, some simple combinations of sources are easily solved; others are not.

One simple combination is two simple sources separated by a distance and vibrating out of phase, one miniature sphere expanding while the other is contracting. The pair is known as a doublet, or dipole, and the radiation of this combination is similar to that of a very small dynamic loudspeaker operating without a baffle. The directivity of a dipole is a figure 8 shape with maximum output along a vector that connects the two sources and minimums to the sides when the observing point is equidistant from the two sources, where the sum of the positive and negative waves cancel each other. While most drivers are dipoles, depending on the enclosure to which they are attached, they may radiate as monopoles, dipoles (or bipoles). If mounted on a finite baffle, and these out of phase waves are allowed to interact, dipole peaks and nulls in the frequency response result. When the rear radiation is absorbed or trapped in a box, the diaphragm becomes a monopole radiator. Bipolar speakers, made by mounting in-phase monopoles (both moving out of or into the box in unison) on opposite sides of a box, are a method of approaching omnidirectional radiation patterns.

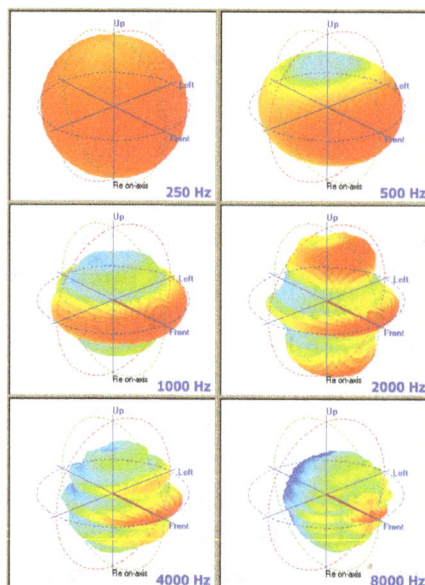

Polar plots of a four-driver industrial columnar public address loudspeaker taken at six frequencies. The pattern is nearly omnidirectional at low frequencies, converging to a wide fan-shaped pattern at 1 kHz, then separating into lobes and getting weaker at higher frequencies

In real life, individual drivers are complex 3D shapes such as cones and domes, and they are placed on a baffle for various reasons. A mathematical expression for the directivity of a complex shape, based on modeling combinations of point sources, is usually not possible, but in the far field, the directivity of a loudspeaker with a circular diaphragm is close to that of a flat circular piston, so it can be used as an illustrative simplification for discussion. As a simple example of the mathematical physics involved, consider the following: the formula for far field directivity of a flat circular piston in an infinite baffle is $p(\theta) = \dfrac{p_0 J_1(k_a \sin \theta)}{k_a \sin \theta}$ where $k_a = \dfrac{2\pi a}{\lambda}$, p_0 is the pressure on axis, a is the piston radius, λ is the wavelength (i.e. $\lambda = \dfrac{c}{f} = \dfrac{\text{speed of sound}}{\text{frequency}}$) θ is the angle off axis and J_1 is the Bessel function of the first kind.

A planar source radiates sound uniformly for low frequencies' wavelengths longer than the dimensions of the planar source, and as frequency increases, the sound from such a source focuses into an increasingly narrower angle. The smaller the driver, the higher the frequency where this narrowing of directivity occurs. Even if the diaphragm is not perfectly circular, this effect occurs such that larger sources are more directive. Several loudspeaker designs approximate this behavior. Most are electrostatic or planar magnetic designs.

Various manufacturers use different driver mounting arrangements to create a specific type of sound field in the space for which they are designed. The resulting radiation patterns may be intended to more closely simulate the way sound is produced by real instruments, or simply create a controlled energy distribution from the input signal (some using this approach are called monitors, as they are useful in checking the signal just recorded in a studio). An example of the first is a room corner system with many small drivers on the surface of a 1/8 sphere. A system design of this type was patented and produced commercially by Professor Amar Bose—the 2201. Later Bose models have deliberately emphasized production of both direct and reflected sound by the loudspeaker itself, regardless of its environment. The designs are controversial in high fidelity circles, but have proven commercially successful. Several other manufacturers' designs follow similar principles.

Directivity is an important issue because it affects the frequency balance of sound a listener hears, and also the interaction of the speaker system with the room and its contents. A very directive (sometimes termed 'beamy') speaker (i.e., on an axis perpendicular to the speaker face) may result in a reverberant field lacking in high frequencies, giving the impression the speaker is deficient in treble even though it measures well on axis (e.g., "flat" across the entire frequency range). Speakers with very wide, or rapidly increasing directivity at high frequencies, can give the impression that there is too much treble (if the listener is on axis) or too little (if the listener is off axis). This is part of the reason why on-axis frequency response measurement is not a complete characterization of the sound of a given loudspeaker.

Other Speaker Designs

While dynamic cone speakers remain the most popular choice, many other speaker technologies exist.

With a Diaphragm

Moving-Iron Loudspeakers

The moving iron speaker was the first type of speaker that was invented. Unlike the newer dynamic (moving coil) design, a moving-iron speaker uses a stationary coil to vibrate a magnetized piece of metal (called the iron, reed, or armature). The metal is either attached to the diaphragm, or is the diaphragm itself. This design was the original loudspeaker design, dating back to the early telephone. Moving iron drivers are inefficient and can only produce a small band of sound. They require large magnets and coils to increase force.

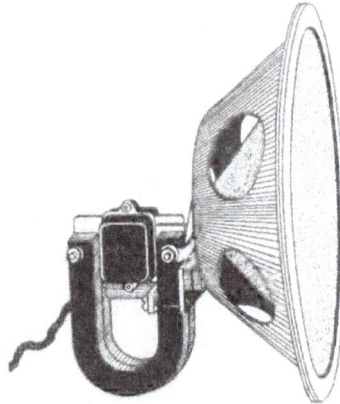

Moving iron speaker

Balanced armature drivers (a type of moving iron driver) use an armature that moves like a see-saw or diving board. Since they are not damped, they are highly efficient, but they also produce strong resonances. They are still used today for high end earphones and hearing aids, where small size and high efficiency are important.

Piezoelectric Speakers

A piezoelectric buzzer. The white ceramic piezoelectric material can be seen fixed to a metal diaphragm.

Piezoelectric speakers are frequently used as beepers in watches and other electronic devices, and are sometimes used as tweeters in less-expensive speaker systems, such as computer speakers and portable radios. Piezoelectric speakers have several advantages over conventional loudspeakers: they are resistant to overloads that would normally destroy most high frequency drivers, and they can be used without a crossover due to their electrical properties. There are also disadvantages: some amplifiers can oscillate when driving capacitive loads like most piezoelectrics, which results in distortion or damage to the amplifier. Additionally, their frequency response, in most cases, is inferior to that of other technologies. This is why they are generally used in single frequency (beeper) or non-critical applications.

Piezoelectric speakers can have extended high frequency output, and this is useful in some specialized circumstances; for instance, sonar applications in which piezoelectric variants are used as both output devices (generating underwater sound) and as input devices (acting as the sensing components of underwater microphones). They have advantages in these applications, not the least of which is simple and solid state construction that resists seawater better than a ribbon or cone based device would.

In 2013, Kyocera introduced piezoelectric ultra-thin medium-size film speakers with only 1 millimeter of thickness and 7 grams of weight for their 55" OLED televisions and they hope the speakers will also be used in PCs and tablets. Besides medium-size, there are also large and small sizes which can all produce relatively the same quality of sound and volume within 180 degrees. The highly responsive speaker material provides better clarity than traditional TV speakers.

Magnetostatic Loudspeakers

Magnetostatic loudspeaker

Instead of a voice coil driving a speaker cone, a magnetostatic speaker uses an array of metal strips bonded to a large film membrane. The magnetic field produced by signal current flowing through the strips interacts with the field of permanent bar magnets mounted behind them. The force produced moves the membrane and so the air in front of it. Typically, these designs are less efficient than conventional moving-coil speakers.

Magnetostrictive Speakers

Magnetostrictive transducers, based on magnetostriction, have been predominantly used as sonar ultrasonic sound wave radiators, but their use has spread also to audio speaker systems. Magnetostrictive speaker drivers have some special advantages: they can provide greater force (with smaller excursions) than other technologies; low excursion can avoid distortions from large excursion as in other designs; the magnetizing coil is stationary and therefore more easily cooled; they are robust because delicate suspensions and voice coils are not required. Magnetostrictive speaker modules have been produced by Fostex and FeONIC and subwoofer drivers have also been produced.

Electrostatic Loudspeakers

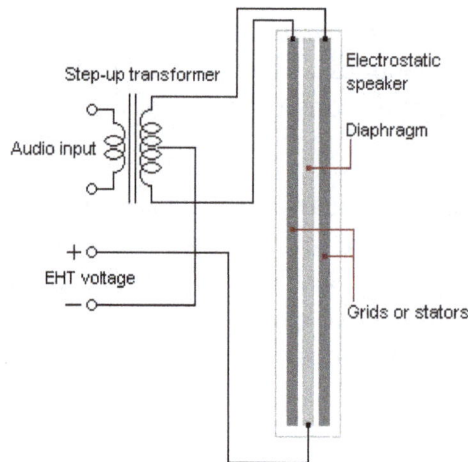

Schematic showing an electrostatic speaker's construction and its connections.
The thickness of the diaphragm and grids has been exaggerated for the purpose of illustration.

Electrostatic loudspeakers use a high voltage electric field (rather than a magnetic field) to drive a thin statically charged membrane. Because they are driven over the entire membrane surface rather than from a small voice coil, they ordinarily provide a more linear and lower-distortion motion than dynamic drivers. They also have a relatively narrow dispersion pattern that can make for precise sound-field positioning. However, their optimum listening area is small and they are not very efficient speakers. They have the disadvantage that the diaphragm excursion is severely limited because of practical construction limitations—the further apart the stators are positioned, the higher the voltage must be to achieve acceptable efficiency. This increases the tendency for electrical arcs as well as increasing the speaker's attraction of dust particles. Arcing remains a potential problem with current technologies, especially when the panels are allowed to collect dust or dirt and are driven with high signal levels.

Electrostatics are inherently dipole radiators and due to the thin flexible membrane are less suited for use in enclosures to reduce low frequency cancellation as with common cone drivers. Due to this and the low excursion capability, full range electrostatic loudspeakers are large by nature, and the bass rolls off at a frequency corresponding to a quarter wavelength of the narrowest panel dimension. To reduce the size of commercial products, they are sometimes used as a high frequency driver in combination with a conventional dynamic driver that handles the bass frequencies effectively.

Electrostatics are usually driven through a step-up transformer that multiplies the voltage swings produced by the power amplifier. This transformer also multiplies the capacitive load that is inherent in electrostatic transducers, which means the effective impedance presented to the power amplifiers varies widely by frequency. A speaker that is nominally 8 ohms may actually present a load of 1 ohm at higher frequencies, which is challenging to some amplifier designs.

Ribbon and Planar Magnetic Loudspeakers

A ribbon speaker consists of a thin metal-film ribbon suspended in a magnetic field. The electrical signal is applied to the ribbon, which moves with it to create the sound. The advantage of a ribbon driver is that the ribbon has very little mass; thus, it can accelerate very quickly, yielding very good high-frequency response. Ribbon loudspeakers are often very fragile—some can be torn by a strong gust of air. Most ribbon tweeters emit sound in a dipole pattern. A few have backings that limit the dipole radiation pattern. Above and below the ends of the more or less rectangular ribbon, there is less audible output due to phase cancellation, but the precise amount of directivity depends on ribbon length. Ribbon designs generally require exceptionally powerful magnets, which makes them costly to manufacture. Ribbons have a very low resistance that most amplifiers cannot drive directly. As a result, a step down transformer is typically used to increase the current through the ribbon. The amplifier "sees" a load that is the ribbon's resistance times the transformer turns ratio squared. The transformer must be carefully designed so that its frequency response and parasitic losses do not degrade the sound, further increasing cost and complication relative to conventional designs.

Planar magnetic speakers (having printed or embedded conductors on a flat diaphragm) are sometimes described as ribbons, but are not truly ribbon speakers. The term planar is generally reserved for speakers with roughly rectangular flat surfaces that radiate in a bipolar (i.e., front and back) manner. Planar magnetic speakers consist of a flexible membrane with a voice coil printed or mounted on it. The current flowing through the coil interacts with the magnetic field of carefully placed magnets on either side of the diaphragm, causing the membrane to vibrate more or less uniformly and without much bending or wrinkling. The driving force covers a large percentage of the membrane surface and reduces resonance problems inherent in coil-driven flat diaphragms.

Bending Wave Loudspeakers

Bending wave transducers use a diaphragm that is intentionally flexible. The rigidity of the material increases from the center to the outside. Short wavelengths radiate primarily from the inner area, while longer waves reach the edge of the speaker. To prevent reflections from the outside back into the center, long waves are absorbed by a surrounding damper. Such transducers can cover a wide frequency range (80 Hz to 35,000 Hz) and have been promoted as being close to an ideal point sound source. This uncommon approach is being taken by only a very few manufacturers, in very different arrangements.

The Ohm Walsh loudspeakers use a unique driver designed by Lincoln Walsh, who had been a radar development engineer in WWII. He became interested in audio equipment design and his last project was a unique, one-way speaker using a single driver. The cone faced down into a sealed, airtight enclosure. Rather than move back-and-forth as conventional speakers do, the cone rippled and created sound in a manner known in RF electronics as a "transmission line". The new speaker created a cylindrical sound field. Lincoln Walsh died before his speaker was released to the public. The Ohm Acoustics firm has produced several loudspeaker models using the Walsh driver design since then. German Physiks, an audio equipment firm in Germany, also produces speakers using this approach.

The German firm, Manger, has designed and produced a bending wave driver that at first glance appears conventional. In fact, the round panel attached to the voice coil bends in a carefully controlled way to produce full range sound. Josef W. Manger was awarded with the "Diesel Medal" for extraordinary developments and inventions by the German institute of inventions.

Flat Panel Loudspeakers

There have been many attempts to reduce the size of speaker systems, or alternatively to make them less obvious. One such attempt was the development of "exciter" transducer coils mounted to flat panels to act as sound sources, most accurately called exciter/panel drivers. These can then be made in a neutral color and hung on walls where they are less noticeable than many speakers, or can be deliberately painted with patterns, in which case they can function decoratively. There are two related problems with flat panel techniques: first, a flat panel is necessarily more flexible than a cone shape in the same material, and therefore moves as a single unit even less, and second, resonances in the panel are difficult to control, leading to considerable distortions. Some progress has been made using such lightweight, rigid, materials such as Styrofoam, and there have been several flat panel systems commercially produced in recent years.

Heil Air Motion Transducers

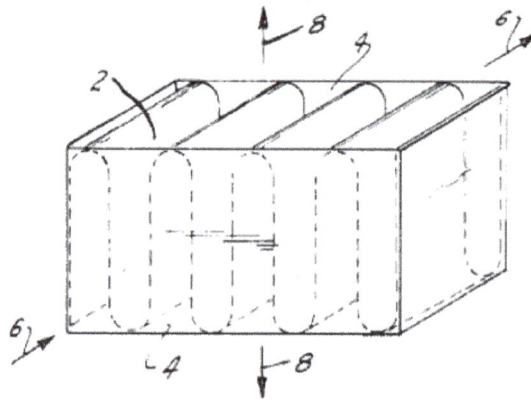

In Heil's air motion transducer, current through the membrane 2 causes it to move left and right in magnetic field 6, moving air in and out along directions 8; barriers 4 prevent air from moving in unintended directions.

Oskar Heil invented the air motion transducer in the 1960s. In this approach, a pleated diaphragm is mounted in a magnetic field and forced to close and open under control of a music signal. Air is forced from between the pleats in accordance with the imposed signal, generating sound. The drivers are less fragile than ribbons and considerably more efficient (and able to produce higher absolute output levels) than ribbon, electrostatic, or planar magnetic tweeter designs. ESS, a California manufacturer, licensed the design, employed Heil, and produced a range of speaker systems using his tweeters during the 1970s and 1980s. Lafayette Radio, a large US retail store chain, also sold speaker systems using such tweeters for a time. There are several manufacturers of these drivers (at least two in Germany—one of which produces a range of high-end professional speakers using tweeters and mid-range drivers based on the technology) and the drivers are increasingly used in professional audio. Martin Logan produces several AMT speakers in the US and GoldenEar Technologies incorporates them in its entire speaker line.

Transparent Ionic Conduction Speaker

In 2013, a research team introduced Transparent ionic conduction speaker which a 2 layers transparent conductive gel and a layer of transparent rubber in between to make high voltage and high actuation work to reproduce good sound quality. The speaker is suitable for robotics, mobile computing and adaptive optics fields.

Without a Diaphragm

Plasma Arc Speakers

Plasma speaker

Plasma arc loudspeakers use electrical plasma as a radiating element. Since plasma has minimal mass, but is charged and therefore can be manipulated by an electric field, the result is a very linear output at frequencies far higher than the audible range. Problems of maintenance and reliability for this approach tend to make it unsuitable for mass market use. In 1978 Alan E. Hill of the Air Force Weapons Laboratory in Albuquerque, NM, designed the Plasmatronics Hill Type I, a tweeter whose plasma was generated from helium gas. This avoided the ozone and nitrous oxide produced by RF decomposition of air in an earlier generation of plasma tweeters made by the pioneering DuKane Corporation, who produced the Ionovac (marketed as the Ionofane in the UK) during the 1950s. Currently, there remain a few manufacturers in Germany who use this design, and a do-it-yourself design has been published and has been available on the Internet.

A less expensive variation on this theme is the use of a flame for the driver, as flames contain ionized (electrically charged) gases.

Thermoacoustic Speakers

In 2008, researchers of Tsinghua University demonstrated a thermoacoustic loudspeaker of carbon nanotube thin film, whose working mechanism is a thermoacoustic effect. Sound frequency electric currents are used to periodically heat the CNT and thus result in sound generation in the surrounding air. The CNT thin film loudspeaker is transparent, stretchable and flexible. In 2013, researchers of Tsinghua University further present a thermoacoustic earphone of carbon nanotube thin yarn and a thermoacoustic surface-mounted device. They are both fully integrated devices and compatible with Si-based semiconducting technology.

Rotary Woofers

A rotary woofer is essentially a fan with blades that constantly change their pitch, allowing them to easily push the air back and forth. Rotary woofers are able to efficiently reproduce infrasound frequencies, which are difficult to impossible to achieve on a traditional speaker with a diaphragm. They are often employed in movie theaters to recreate rumbling bass effects, such as explosions.

New Technologies

Digital Speakers

Digital speakers have been the subject of experiments performed by Bell Labs as far back as the 1920s. The design is simple; each bit controls a driver, which is either fully 'on' or 'off'. Problems with this design have led manufacturers to abandon it as impractical for the present. First, for a reasonable number of bits (required for adequate sound reproduction quality), the physical size of a speaker system becomes very large. Secondly, due to inherent analog digital conversion problems, the effect of aliasing is unavoidable, so that the audio output is "reflected" at equal amplitude in the frequency domain, on the other side of the sampling frequency, causing an unacceptably high level of ultrasonics to accompany the desired output. No workable scheme has been found to adequately deal with this.

The term "digital" or "digital-ready" is often used for marketing purposes on speakers or headphones, but these systems are not digital in the sense described above. Rather, they are conventional speakers that can be used with digital sound sources (e.g., optical media, MP3 players, etc.), as can any conventional speaker.

Microphone

Microphones are used in many applications such as telephones, hearing aids, public address systems for concert halls and public events, motion picture production, live and recorded audio engineering, sound recording, two-way radios, megaphones, radio and television broadcasting, and in computers for recording voice, speech recognition, VoIP, and for non-acoustic purposes such as ultrasonic sensors or knock sensors.

Shure Brothers microphone, model 55s, Multi-Impedance "Small Unidyne" Dynamic from 1951

A Sennheiser dynamic microphone

A microphone, colloquially nicknamed mic or mike (/□ma□k/), is a transducer that converts sound into an electrical signal.

Several different types of microphone are in use, which employ different methods to convert the air pressure variations of a sound wave to an electrical signal. The most common are the dynamic microphone, which uses a coil of wire suspended in a magnetic field; the condenser microphone, which uses the vibrating diaphragm as a capacitor plate, and the piezoelectric microphone, which uses a crystal of piezoelectric material. Microphones typically need to be connected to a preamplifier before the signal can be recorded or reproduced.

History

In order to speak to larger groups of people, a need arose to increase the volume of the human voice. The earliest devices used to achieve this were acoustic megaphones. Some of the first examples, from fifth century BC Greece, were theater masks with horn-shaped mouth openings that acoustically amplified the voice of actors in amphitheatres. In 1665, the English physicist Robert Hooke was the first to experiment with a medium other than air with the invention of the "lovers' telephone" made of stretched wire with a cup attached at each end.

German inventor Johann Philipp Reis designed an early sound transmitter that used a metallic strip attached to a vibrating membrane that would produce intermittent current. Better results were achieved with the "liquid transmitter" design in Scottish-American Alexander Graham Bell's telephone of 1876 – the diaphragm was attached to a conductive rod in an acid solution. These systems, however, gave a very poor sound quality.

David Edward Hughes invented a carbon microphone in the 1870s.

The first microphone that enabled proper voice telephony was the (loose-contact) carbon microphone. This was independently developed by David Edward Hughes in England and Emile Berliner and Thomas Edison in the US. Although Edison was awarded the first patent (after a long legal dispute) in mid-1877, Hughes had demonstrated his working device in front of many witnesses some years earlier, and most historians credit him with its invention. The carbon microphone is the direct prototype of today's microphones and was critical in the development of telephony, broadcasting and the recording industries. Thomas Edison refined the carbon microphone into his carbon-button transmitter of 1886. This microphone was employed at the first ever radio broadcast, a performance at the New York Metropolitan Opera House in 1910.

Jack Brown interviews Humphrey Bogart and Lauren Bacall for broadcast to troops overseas during World War II.

In 1916, E.C. Wente of Western Electric developed the next breakthrough with the first condenser microphone. In 1923, the first practical moving coil microphone was built. "The Marconi Skykes" or "magnetophon", developed by Captain H. J. Round, was the standard for BBC studios in London. This was improved in 1930 by Alan Blumlein and Herbert Holman who released the HB1A and was the best standard of the day.

Also in 1923, the ribbon microphone was introduced, another electromagnetic type, believed to have been developed by Harry F. Olson, who essentially reverse-engineered a ribbon speaker. Over the years these microphones were developed by several companies, most notably RCA that made large advancements in pattern control, to give the microphone directionality. With television and film technology booming there was demand for high fidelity microphones and greater directionality. Electro-Voice responded with their Academy Award-winning shotgun microphone in 1963.

During the second half of 20th century development advanced quickly with the Shure Brothers bringing out the SM58 and SM57. Digital was pioneered by Milab in 1999 with the DM-1001. The latest research developments include the use of fibre optics, lasers and interferometers.

Components

Electronic symbol for a microphone

The sensitive transducer element of a microphone is called its *element* or *capsule*. Sound is first converted to mechanical motion by means of a diaphragm, the motion of which is then converted to an electrical signal. A complete microphone also includes a housing, some means of bringing the signal from the element to other equipment, and often an electronic circuit to adapt the output of the capsule to the equipment being driven. A wireless microphone contains a radio transmitter.

Varieties

Microphones are categorized by their transducer principle, such as condenser, dynamic, etc., and by their directional characteristics. Sometimes other characteristics such as diaphragm size, intended use or orientation of the principal sound input to the principal axis (end- or side-address) of the microphone are used to describe the microphone.

Condenser

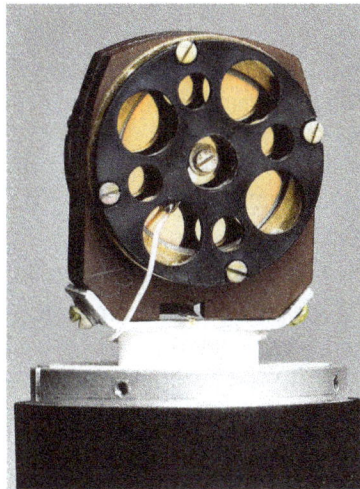

Inside the Oktava 319 condenser microphone

The condenser microphone, invented at Western Electric in 1916 by E. C. Wente, is also called a capacitor microphone or electrostatic microphone—capacitors were historically called condensers. Here, the diaphragm acts as one plate of a capacitor, and the vibrations produce changes in the distance between the plates. There are two types, depending on the method of extracting the audio signal from the transducer: DC-biased microphones, and radio frequency (RF) or high frequency (HF) condenser microphones. With a DC-biased microphone, the plates are biased with a fixed charge (Q). The voltage maintained across the capacitor plates changes with the vibrations in the air, according to the capacitance equation ($C = \frac{Q}{V}$), where Q = charge in coulombs, C = capacitance in farads and V = potential difference in volts. The capacitance of the plates is inversely proportional to the distance between them for a parallel-plate capacitor. The assembly of fixed and movable plates is called an "element" or "capsule".

A nearly constant charge is maintained on the capacitor. As the capacitance changes, the charge across the capacitor does change very slightly, but at audible frequencies it is sensibly constant. The capacitance of the capsule (around 5 to 100 pF) and the value of the bias resistor (100 MΩ to tens of GΩ) form a filter that is high-pass for the audio signal, and low-pass for the bias voltage. The time constant of an RC circuit equals the product of the resistance and capacitance.

Within the time-frame of the capacitance change (as much as 50 ms at 20 Hz audio signal), the charge is practically constant and the voltage across the capacitor changes instantaneously to reflect the change in capacitance. The voltage across the capacitor varies above and below the bias voltage. The voltage difference between the bias and the capacitor is seen across the series resistor. The voltage across the resistor is amplified for performance or recording. In most cases, the electronics in the microphone itself contribute no voltage gain as the voltage differential is quite significant, up to several volts for high sound levels. Since this is a very high impedance circuit, current gain only is usually needed, with the voltage remaining constant.

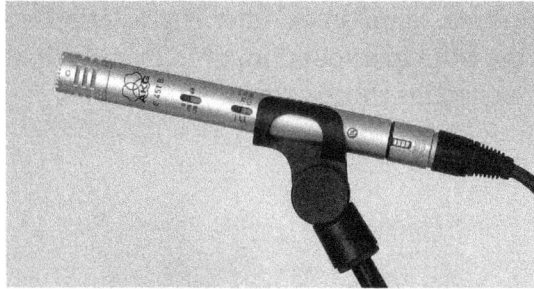
AKG C451B small-diaphragm condenser microphone

RF condenser microphones use a comparatively low RF voltage, generated by a low-noise oscillator. The signal from the oscillator may either be amplitude modulated by the capacitance changes produced by the sound waves moving the capsule diaphragm, or the capsule may be part of a resonant circuit that modulates the frequency of the oscillator signal. Demodulation yields a low-noise audio frequency signal with a very low source impedance. The absence of a high bias voltage permits the use of a diaphragm with looser tension, which may be used to achieve wider frequency response due to higher compliance. The RF biasing process results in a lower electrical impedance capsule, a useful by-product of which is that RF condenser microphones can be operated in damp weather conditions that could create problems in DC-biased microphones with contaminated insulating surfaces. The Sennheiser "MKH" series of microphones use the RF biasing technique.

Condenser microphones span the range from telephone transmitters through inexpensive karaoke microphones to high-fidelity recording microphones. They generally produce a high-quality audio signal and are now the popular choice in laboratory and recording studio applications. The inherent suitability of this technology is due to the very small mass that must be moved by the incident sound wave, unlike other microphone types that require the sound wave to do more work. They require a power source, provided either via microphone inputs on equipment as phantom power or from a small battery. Power is necessary for establishing the capacitor plate voltage, and is also needed to power the microphone electronics (impedance conversion in the case of electret and DC-polarized microphones, demodulation or detection in the case of RF/HF microphones). Condenser microphones are also available with two diaphragms that can be electrically connected to provide a range of polar patterns, such as cardioid, omnidirectional, and figure-eight. It is also possible to vary the pattern continuously with some microphones, for example the Røde NT2000 or CAD M179.

A valve microphone is a condenser microphone that uses a vacuum tube (valve) amplifier. They remain popular with enthusiasts of tube sound.

Electret Condenser

An electret microphone is a type of capacitor microphone invented by Gerhard Sessler and Jim West at Bell laboratories in 1962. The externally applied charge described above under condenser microphones is replaced by a permanent charge in an electret material. An electret is a ferroelectric material that has been permanently electrically charged or *polarized*. The name comes from *electr*ostatic and magn*et*; a static charge is embedded in an electret by alignment of the static charges in the material, much the way a magnet is made by aligning the magnetic domains in a piece of iron.

Due to their good performance and ease of manufacture, hence low cost, the vast majority of microphones made today are electret microphones; a semiconductor manufacturer estimates annual production at over one billion units. Nearly all cell-phone, computer, PDA and headset microphones are electret types. They are used in many applications, from high-quality recording and lavalier use to built-in microphones in small sound recording devices and telephones. Though electret microphones were once considered low quality, the best ones can now rival traditional condenser microphones in every respect and can even offer the long-term stability and ultra-flat response needed for a measurement microphone. Unlike other capacitor microphones, they require no polarizing voltage, but often contain an integrated preamplifier that does require power (often incorrectly called polarizing power or bias). This preamplifier is frequently phantom powered in sound reinforcement and studio applications. Monophonic microphones designed for personal computer (PC) use, sometimes called multimedia microphones, use a 3.5 mm plug as usually used, without power, for stereo; the ring, instead of carrying the signal for a second channel, carries power via a resistor from (normally) a 5 V supply in the computer. Stereophonic microphones use the same connector; there is no obvious way to determine which standard is used by equipment and microphones.

Only the best electret microphones rival good DC-polarized units in terms of noise level and quality; electret microphones lend themselves to inexpensive mass-production, while inherently expensive non-electret condenser microphones are made to higher quality.

Dynamic

Patti Smith singing into a Shure SM58 (dynamic cardioid type) microphone

The dynamic microphone (also known as the moving-coil microphone) works via electromagnetic induction. They are robust, relatively inexpensive and resistant to moisture. This, coupled with their potentially high gain before feedback, makes them ideal for on-stage use.

Dynamic microphones use the same dynamic principle as in a loudspeaker, only reversed. A small movable induction coil, positioned in the magnetic field of a permanent magnet, is attached to the diaphragm. When sound enters through the windscreen of the microphone, the sound wave moves the diaphragm. When the diaphragm vibrates, the coil moves in the magnetic field, producing a varying current in the coil through electromagnetic induction. A single dynamic membrane does not respond linearly to all audio frequencies. For this reason some microphones utilize multiple membranes for the different parts of the audio spectrum and then combine the resulting signals. Combining the multiple signals correctly is difficult and designs that do this are rare and tend to be expensive. On the other hand there are several designs that are more specifically aimed towards isolated parts of the audio spectrum. The AKG D 112, for example, is designed for bass response rather than treble. In audio engineering several kinds of microphones are often used at the same time to get the best results.

Ribbon

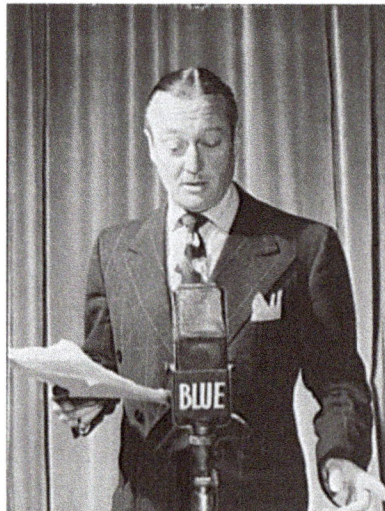

Edmund Lowe using a ribbon microphone

Ribbon microphones use a thin, usually corrugated metal ribbon suspended in a magnetic field. The ribbon is electrically connected to the microphone's output, and its vibration within the magnetic field generates the electrical signal. Ribbon microphones are similar to moving coil microphones in the sense that both produce sound by means of magnetic induction. Basic ribbon microphones detect sound in a bi-directional (also called figure-eight, as in the diagram below) pattern because the ribbon is open on both sides. Also, because the ribbon is much less mass it responds to the air velocity rather than the sound pressure. Though the symmetrical front and rear pickup can be a nuisance in normal stereo recording, the high side rejection can be used to advantage by positioning a ribbon microphone horizontally, for example above cymbals, so that the rear lobe picks up only sound from the cymbals. Crossed figure 8, or Blumlein pair, stereo recording is gaining in popularity, and the figure-eight response of a ribbon microphone is ideal for that application.

Other directional patterns are produced by enclosing one side of the ribbon in an acoustic trap or baffle, allowing sound to reach only one side. The classic RCA Type 77-DX microphone has several externally adjustable positions of the internal baffle, allowing the selection of several response patterns ranging from "figure-eight" to "unidirectional". Such older ribbon microphones, some of which still provide high quality sound reproduction, were once valued for this reason, but a good low-frequency response could only be obtained when the ribbon was suspended very loosely, which made them relatively fragile. Modern ribbon materials, including new nanomaterials have now been introduced that eliminate those concerns, and even improve the effective dynamic range of ribbon microphones at low frequencies. Protective wind screens can reduce the danger of damaging a vintage ribbon, and also reduce plosive artifacts in the recording. Properly designed wind screens produce negligible treble attenuation. In common with other classes of dynamic microphone, ribbon microphones don't require phantom power; in fact, this voltage can damage some older ribbon microphones. Some new modern ribbon microphone designs incorporate a preamplifier and, therefore, do require phantom power, and circuits of modern passive ribbon microphones, *i.e.*, those without the aforementioned preamplifier, are specifically designed to resist damage to the ribbon and transformer by phantom power. Also there are new ribbon materials available that are immune to wind blasts and phantom power.

Carbon

A carbon microphone, also known as a carbon button microphone (or sometimes just a button microphone), uses a capsule or button containing carbon granules pressed between two metal plates like the Berliner and Edison microphones. A voltage is applied across the metal plates, causing a small current to flow through the carbon. One of the plates, the diaphragm, vibrates in sympathy with incident sound waves, applying a varying pressure to the carbon. The changing pressure deforms the granules, causing the contact area between each pair of adjacent granules to change, and this causes the electrical resistance of the mass of granules to change. The changes in resistance cause a corresponding change in the current flowing through the microphone, producing the electrical signal. Carbon microphones were once commonly used in telephones; they have extremely low-quality sound reproduction and a very limited frequency response range, but are very robust devices. The Boudet microphone, which used relatively large carbon balls, was similar to the granule carbon button microphones.

Unlike other microphone types, the carbon microphone can also be used as a type of amplifier, using a small amount of sound energy to control a larger amount of electrical energy. Carbon microphones found use as early telephone repeaters, making long distance phone calls possible in the era before vacuum tubes. These repeaters worked by mechanically coupling a magnetic telephone receiver to a carbon microphone: the faint signal from the receiver was transferred to the microphone, where it modulated a stronger electric current, producing a stronger electrical signal to send down the line. One illustration of this amplifier effect was the oscillation caused by feedback, resulting in an audible squeal from the old "candlestick" telephone if its earphone was placed near the carbon microphone.

Piezoelectric

A crystal microphone or piezo microphone uses the phenomenon of piezoelectricity—the ability of some materials to produce a voltage when subjected to pressure—to convert vibrations into an electrical signal. An example of this is potassium sodium tartrate, which is a piezoelectric crystal that works as a transducer, both as a microphone and as a slimline loudspeaker component. Crystal microphones were once commonly supplied with vacuum tube (valve) equipment, such as domestic tape recorders. Their high output impedance matched the high input impedance (typically about 10 megohms) of the vacuum tube input stage well. They were difficult to match to early transistor equipment, and were quickly supplanted by dynamic microphones for a time, and later small electret condenser devices. The high impedance of the crystal microphone made it very susceptible to handling noise, both from the microphone itself and from the connecting cable.

Piezoelectric transducers are often used as contact microphones to amplify sound from acoustic musical instruments, to sense drum hits, for triggering electronic samples, and to record sound in challenging environments, such as underwater under high pressure. Saddle-mounted pickups on acoustic guitars are generally piezoelectric devices that contact the strings passing over the saddle. This type of microphone is different from magnetic coil pickups commonly visible on typical electric guitars, which use magnetic induction, rather than mechanical coupling, to pick up vibration.

Fiber Optic

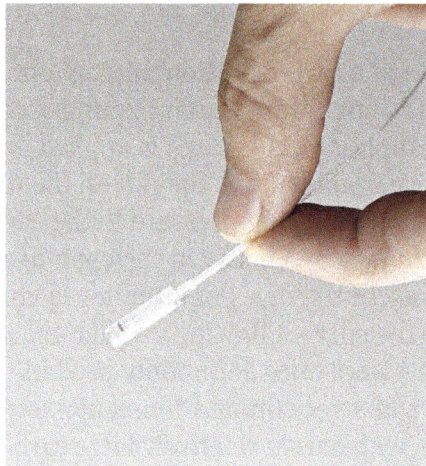

The Optoacoustics 1140 fiber optic microphone

A fiber optic microphone converts acoustic waves into electrical signals by sensing changes in light intensity, instead of sensing changes in capacitance or magnetic fields as with conventional microphones.

During operation, light from a laser source travels through an optical fiber to illuminate the surface of a reflective diaphragm. Sound vibrations of the diaphragm modulate the intensity of light reflecting off the diaphragm in a specific direction. The modulated light is then transmitted over a second optical fiber to a photo detector, which transforms the intensity-modulated light into analog or digital audio for transmission or recording. Fiber optic microphones possess high dynamic and frequency range, similar to the best high fidelity conventional microphones.

Fiber optic microphones do not react to or influence any electrical, magnetic, electrostatic or radioactive fields (this is called EMI/RFI immunity). The fiber optic microphone design is therefore ideal for use in areas where conventional microphones are ineffective or dangerous, such as inside industrial turbines or in magnetic resonance imaging (MRI) equipment environments.

Fiber optic microphones are robust, resistant to environmental changes in heat and moisture, and can be produced for any directionality or impedance matching. The distance between the microphone's light source and its photo detector may be up to several kilometers without need for any preamplifier or other electrical device, making fiber optic microphones suitable for industrial and surveillance acoustic monitoring.

Fiber optic microphones are used in very specific application areas such as for infrasound monitoring and noise-canceling. They have proven especially useful in medical applications, such as allowing radiologists, staff and patients within the powerful and noisy magnetic field to converse normally, inside the MRI suites as well as in remote control rooms. Other uses include industrial equipment monitoring and audio calibration and measurement, high-fidelity recording and law enforcement.

Laser

Laser microphones are often portrayed in movies as spy gadgets, because they can be used to pick up sound at a distance from the microphone equipment. A laser beam is aimed at the surface of a window or other plane surface that is affected by sound. The vibrations of this surface change the angle at which the beam is reflected, and the motion of the laser spot from the returning beam is detected and converted to an audio signal.

In a more robust and expensive implementation, the returned light is split and fed to an interferometer, which detects movement of the surface by changes in the optical path length of the reflected beam. The former implementation is a tabletop experiment; the latter requires an extremely stable laser and precise optics.

A new type of laser microphone is a device that uses a laser beam and smoke or vapor to detect sound vibrations in free air. On 25 August 2009, U.S. patent 7,580,533 issued for a Particulate Flow Detection Microphone based on a laser-photocell pair with a moving stream of smoke or vapor in the laser beam's path. Sound pressure waves cause disturbances in the smoke that in turn cause variations in the amount of laser light reaching the photo detector. A prototype of the device was demonstrated at the 127th Audio Engineering Society convention in New York City from 9 through 12 October 2009.

Liquid

Early microphones did not produce intelligible speech, until Alexander Graham Bell made improvements including a variable-resistance microphone/transmitter. Bell's liquid transmitter consisted of a metal cup filled with water with a small amount of sulfuric acid added. A sound wave caused the diaphragm to move, forcing a needle to move up and down in the water. The electrical resistance between the wire and the cup was then inversely proportional to the size of the water meniscus around the submerged needle. Elisha Gray filed a caveat for a version using a brass rod instead of the needle. Other minor variations and improvements were made to the liquid microphone by Majoranna, Chambers, Vanni, Sykes, and Elisha Gray, and one version was patented by Reginald Fessenden in 1903. These were the first working microphones, but they were not practical for commercial application. The famous first phone conversation between Bell and Watson took place using a liquid microphone.

MEMS

The MEMS (MicroElectrical-Mechanical System) microphone is also called a microphone chip or silicon microphone. A pressure-sensitive diaphragm is etched directly into a silicon wafer by MEMS processing techniques, and is usually accompanied with integrated preamplifier. Most MEMS microphones are variants of the condenser microphone design. Digital MEMS microphones have built in analog-to-digital converter (ADC) circuits on the same CMOS chip making the chip a digital microphone and so more readily integrated with modern digital products. Major manufacturers producing MEMS silicon microphones are Wolfson Microelectronics (WM7xxx) now Cirrus Logic, InvenSense (product line sold by Analog Devices), Akustica (AKU200x), Infineon (SMM310 product), Knowles Electronics, Memstech (MSMx), NXP Semiconductors (division bought by Knowles), Sonion MEMS, Vesper, AAC Acoustic Technologies, and Omron.

More recently, there has been increased interest and research into making piezoelectric MEMS microphones which are a significant architectural and material change from existing condenser style MEMS designs.

Speakers as Microphones

A loudspeaker, a transducer that turns an electrical signal into sound waves, is the functional opposite of a microphone. Since a conventional speaker is constructed much like a dynamic microphone (with a diaphragm, coil and magnet), speakers can actually work "in reverse" as microphones. The resulting signal typically offers reduced quality including limited high-end frequency response and poor sensitivity. In practical use, speakers are sometimes used as microphones in applications where high quality and sensitivity are not needed such as intercoms, walkie-talkies or video game voice chat peripherals, or when conventional microphones are in short supply.

However, there is at least one practical application that exploits those weaknesses: the use of a medium-size woofer placed closely in front of a "kick drum" (bass drum) in a drum set to act as a microphone. A commercial product example is the Yamaha Subkick, a 6.5-inch (170 mm) woofer shock-mounted into a 10" drum shell used in front of kick drums. Since a relatively massive membrane is unable to transduce high frequencies while being capable of tolerating strong low-frequency transients, the speaker is often ideal for picking up the kick drum while reducing bleed from the nearby cymbals and snare drums. Less commonly, microphones themselves can be used as speakers, but due to their low power handling and small transducer sizes, a tweeter is the most practical application. One instance of such an application was the STC microphone-derived 4001 super-tweeter, which was successfully used in a number of high quality loudspeaker systems from the late 1960s to the mid-70s.

Capsule Design and Directivity

The inner elements of a microphone are the primary source of differences in directivity. A pressure microphone uses a diaphragm between a fixed internal volume of air and the environment, and responds uniformly to pressure from all directions, so it is said to be omnidirectional. A pressure-gradient microphone uses a diaphragm that is at least partially open on both sides. The pressure difference between the two sides produces its directional characteristics. Other elements such as the external shape of the microphone and external devices such as interference tubes can also alter a microphone's directional response. A pure pressure-gradient microphone is equally sensitive to sounds arriving from front or back, but insensitive to sounds arriving from the side because sound arriving at the front and back at the same time creates no gradient between the two. The characteristic directional pattern of a pure pressure-gradient microphone is like a figure-8. Other polar patterns are derived by creating a capsule that combines these two effects in different ways. The cardioid, for instance, features a partially closed backside, so its response is a combination of pressure and pressure-gradient characteristics.

Polar Patterns

Microphone facing top of page in diagram, parallel to page:

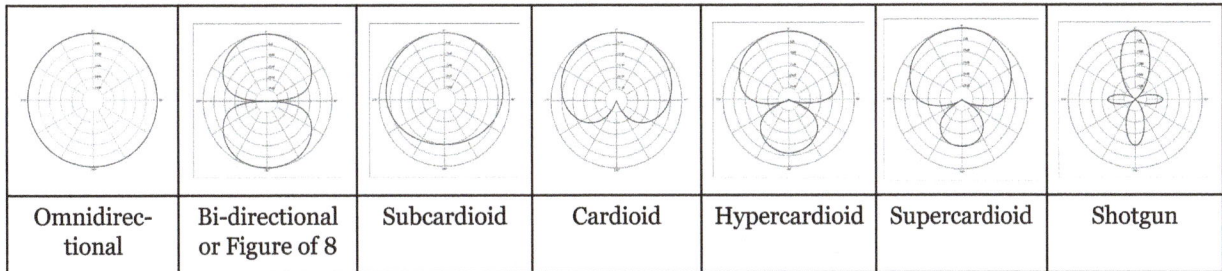

Omnidirec-tional	Bi-directional or Figure of 8	Subcardioid	Cardioid	Hypercardioid	Supercardioid	Shotgun

A microphone's directionality or polar pattern indicates how sensitive it is to sounds arriving at different angles about its central axis. The polar patterns illustrated above represent the locus of points that produce the same signal level output in the microphone if a given sound pressure level (SPL) is generated from that point. How the physical body of the microphone is oriented relative to the diagrams depends on the microphone design. For large-membrane microphones such as in the Oktava (pictured above), the upward direction in the polar diagram is usually perpendicular to the microphone body, commonly known as "side fire" or "side address". For small diaphragm microphones such as the Shure (also pictured above), it usually extends from the axis of the microphone commonly known as "end fire" or "top/end address".

Some microphone designs combine several principles in creating the desired polar pattern. This ranges from shielding (meaning diffraction/dissipation/absorption) by the housing itself to electronically combining dual membranes.

Omnidirectional

An omnidirectional (or nondirectional) microphone's response is generally considered to be a perfect sphere in three dimensions. In the real world, this is not the case. As with directional microphones, the polar pattern for an "omnidirectional" microphone is a function of frequency. The body of the microphone is not infinitely small and, as a consequence, it tends to get in its own way with respect to sounds arriving from the rear, causing a slight flattening of the polar response. This flattening increases as the diameter of the microphone (assuming it's cylindrical) reaches the wavelength of the frequency in question. Therefore, the smallest diameter microphone gives the best omnidirectional characteristics at high frequencies.

The wavelength of sound at 10 kHz is 1.4" (3.5 cm). The smallest measuring microphones are often 1/4" (6 mm) in diameter, which practically eliminates directionality even up to the highest frequencies. Omnidirectional microphones, unlike cardioids, do not employ resonant cavities as delays, and so can be considered the "purest" microphones in terms of low coloration; they add very little to the original sound. Being pressure-sensitive they can also have a very flat low-frequency response down to 20 Hz or below. Pressure-sensitive microphones also respond much less to wind noise and plosives than directional (velocity sensitive) microphones.

An example of a nondirectional microphone is the round black *eight ball.*

Unidirectional

A unidirectional microphone is primarily sensitive to sounds from only one direction. The diagram above illustrates a number of these patterns. The microphone faces upwards in each diagram. The sound intensity for a particular frequency is plotted for angles radially from 0 to 360°. (Professional diagrams show these scales and include multiple plots at different frequencies. The diagrams given here provide only an overview of typical pattern shapes, and their names.)

Cardioid, Hypercardioid, Supercardioid, Subcardioid

University Sound US664A dynamic supercardioid microphone

The most common unidirectional microphone is a cardioid microphone, so named because the sensitivity pattern is "heart-shaped", i.e. a cardioid. The cardioid family of microphones are commonly used as vocal or speech microphones, since they are good at rejecting sounds from other directions. In three dimensions, the cardioid is shaped like an apple centred around the microphone which is the "stem" of the apple. The cardioid response reduces pickup from the side and rear, helping to avoid feedback from the monitors. Since these directional transducer microphones achieve their patterns by sensing pressure gradient, putting them very close to the sound source (at distances of a few centimeters) results in a bass boost due to the increased gradient. This is known as the proximity effect. The SM58 has been the most commonly used microphone for live vocals for more than 50 years demonstrating the importance and popularity of cardioid mics.

A cardioid microphone is effectively a superposition of an omnidirectional and a figure-8 microphone; for sound waves coming from the back, the negative signal from the figure-8 cancels the positive signal from the omnidirectional element, whereas for sound waves coming from the front, the two add to each other. A hyper-cardioid microphone is similar, but with a slightly larger figure-8 contribution leading to a tighter area of front sensitivity and a smaller lobe of rear sensitivity. A super-cardioid microphone is similar to a hyper-cardioid, except there is more front pickup and less rear pickup. While any pattern between omni and figure 8 is possible by adjusting their mix, common definitions state that a hypercardioid is produced by combining them at a 3:1 ratio, producing nulls at 109.5°, while supercardioid is produced with about a 5:3 ratio, with nulls at 126.9°. The sub-cardioid microphone has no null points. It is produced with about 7:3 ratio with 3-10 dB level between the front and back pickup.

Bi-directional

"Figure 8" or bi-directional microphones receive sound equally from both the front and back of the element. Most ribbon microphones are of this pattern. In principle they do not respond to sound pressure at all, only to the *change* in pressure between front and back; since sound arriving from the side reaches front and back equally there is no difference in pressure and therefore no sensitivity to sound from that direction. In more mathematical terms, while omnidirectional microphones are scalar transducers responding to pressure from any direction, bi-directional microphones are vector transducers responding to the gradient along an axis normal to the plane of the diaphragm. This also has the effect of inverting the output polarity for sounds arriving from the back side.

Shotgun and Parabolic

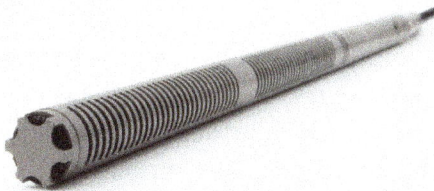

An Audio-Technica shotgun microphone

The interference tube of a shotgun microphone. The capsule is at the base of the tube

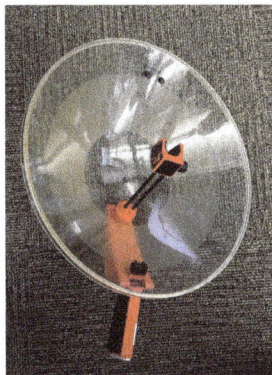

A Sony parabolic reflector, without a microphone. The microphone would face the reflector surface and sound captured by the reflector would bounce towards the microphone.

Shotgun microphones are the most highly directional of simple first-order unidirectional types. At low frequencies they have the classic polar response of a hypercardioid but at medium and higher frequencies an interference tube gives them an increased forward response. This is achieved by a process of cancellation of off-axis waves entering the longitudinal array of slots. A consequence of this technique is the presence of some rear lobes that vary in level and angle with frequency, and can cause some coloration effects. Due to the narrowness of their forward sensitivity, shotgun microphones are commonly used on television and film sets, in stadiums, and for field recording of wildlife.

Boundary or "PZM"

Several approaches have been developed for effectively using a microphone in less-than-ideal acoustic spaces, which often suffer from excessive reflections from one or more of the surfaces (boundaries) that make up the space. If the microphone is placed in, or very close to, one of these boundaries, the reflections from that surface have the same timing as the direct sound, thus giving the microphone a hemispherical polar pattern and improved intelligibility. Initially this was done by placing an ordinary microphone adjacent to the surface, sometimes in a block of acoustically transparent foam. Sound engineers Ed Long and Ron Wickersham developed the concept of placing the diaphragm parallel to and facing the boundary. While the patent has expired, "Pressure Zone Microphone" and "PZM" are still active trademarks of Crown International, and the generic term "boundary microphone" is preferred. While a boundary microphone was initially implemented using an omnidirectional element, it is also possible to mount a directional microphone close enough to the surface to gain some of the benefits of this technique while retaining the directional properties of the element. Crown's trademark on this approach is "Phase Coherent Cardioid" or "PCC," but there are other makers who employ this technique as well.

Application-specific Designs

A lavalier microphone is made for hands-free operation. These small microphones are worn on the body. Originally, they were held in place with a lanyard worn around the neck, but more often they are fastened to clothing with a clip, pin, tape or magnet. The lavalier cord may be hidden by clothes and either run to an RF transmitter in a pocket or clipped to a belt (for mobile use), or run directly to the mixer (for stationary applications).

A wireless microphone transmits the audio as a radio or optical signal rather than via a cable. It usually sends its signal using a small FM radio transmitter to a nearby receiver connected to the sound system, but it can also use infrared waves if the transmitter and receiver are within sight of each other.

A contact microphone picks up vibrations directly from a solid surface or object, as opposed to sound vibrations carried through air. One use for this is to detect sounds of a very low level, such as those from small objects or insects. The microphone commonly consists of a magnetic (moving coil) transducer, contact plate and contact pin. The contact plate is placed directly on the vibrating part of a musical instrument or other surface, and the contact pin transfers vibrations to the coil. Contact microphones have been used to pick up the sound of a snail's heartbeat and the footsteps of ants. A portable version of this microphone has recently been developed. A throat microphone is a variant of the contact microphone that picks up speech directly from a person's throat, which it is strapped to. This lets the device be used in areas with ambient sounds that would otherwise make the speaker inaudible.

A parabolic microphone uses a parabolic reflector to collect and focus sound waves onto a microphone receiver, in much the same way that a parabolic antenna (e.g. satellite dish) does with radio waves. Typical uses of this microphone, which has unusually focused front sensitivity and can pick up sounds from many meters away, include nature recording, outdoor sporting events, eavesdropping, law enforcement, and even espionage. Parabolic microphones are not typically used for standard recording applications, because they tend to have poor low-frequency response as a side effect of their design.

A stereo microphone integrates two microphones in one unit to produce a stereophonic signal. A stereo microphone is often used for broadcast applications or field recording where it would be impractical to configure two separate condenser microphones in a classic X-Y configuration for stereophonic recording. Some such microphones have an adjustable angle of coverage between the two channels.

A noise-canceling microphone is a highly directional design intended for noisy environments. One such use is in aircraft cockpits where they are normally installed as boom microphones on headsets. Another use is in live event support on loud concert stages for vocalists involved with live performances. Many noise-canceling microphones combine signals received from two diaphragms that are in opposite electrical polarity or are processed electronically. In dual diaphragm designs, the main diaphragm is mounted closest to the intended source and the second is positioned farther away from the source so that it can pick up environmental sounds to be subtracted from the main diaphragm's signal. After the two signals have been combined, sounds other than the intended source are greatly reduced, substantially increasing intelligibility. Other noise-canceling designs use one diaphragm that is affected by ports open to the sides and rear of the microphone, with the sum being a 16 dB rejection of sounds that are farther away. One noise-canceling headset design using a single diaphragm has been used prominently by vocal artists such as Garth Brooks and Janet Jackson. A few noise-canceling microphones are throat microphones.

Powering

Microphones containing active circuitry, such as most condenser microphones, require power to operate the active components. The first of these used vacuum-tube circuits with a separate power supply unit, using a multi-pin cable and connector. With the advent of solid-state amplification, the power requirements were greatly reduced and it became practical to use the same cable conductors and connector for audio and power. During the 1960s several powering methods were developed, mainly in Europe. The two dominant methods were initially defined in German DIN 45595 as de:Tonaderspeisung or T-power and DIN 45596 for phantom power. Since the 1980s, phantom power has become much more common, because the same input may be used for both powered and unpowered microphones. In consumer electronics such as DSLRs and camcorders, "plug-in power" is more common, for microphones using a 3.5 mm phone plug connector. Phantom, T-power and plug-in power are described in international standard IEC 61938.

Connectors

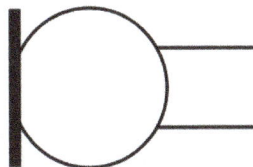

Electronic symbol for a microphone

The most common connectors used by microphones are:

- Male XLR connector on professional microphones

- ¼ inch (sometimes referred to as 6.3 mm) phone connector on less expensive musician's microphones, using an unbalanced 1/4 inch (6.3 mm) TS phone connector. Harmonica microphones commonly use a high impedance 1/4 inch (6.3 mm) TS connection to be run through guitar amplifiers.

- 3.5 mm (sometimes referred to as 1/8 inch mini) stereo (sometimes wired as mono) mini phone plug on prosumer camera, recorder and computer microphones.

A microphone with a USB connector, made by Blue Microphones

Some microphones use other connectors, such as a 5-pin XLR, or mini XLR for connection to portable equipment. Some lavalier (or "lapel", from the days of attaching the microphone to the news reporters suit lapel) microphones use a proprietary connector for connection to a wireless transmitter, such as a radio pack. Since 2005, professional-quality microphones with USB connections have begun to appear, designed for direct recording into computer-based software.

Impedance-matching

Microphones have an electrical characteristic called impedance, measured in ohms (Ω), that depends on the design. In passive microphones, this value describes the electrical resistance of the magnet coil (or similar mechanism). In active microphones, this value describes the output resistance of the amplifier circuitry. Typically, the *rated impedance* is stated. Low impedance is considered under 600 Ω. Medium impedance is considered between 600 Ω and 10 kΩ. High impedance is above 10 kΩ. Owing to their built-in amplifier, condenser microphones typically have an output impedance between 50 and 200 Ω.

The output of a given microphone delivers the same power whether it is low or high impedance. If a microphone is made in high and low impedance versions, the high impedance version has a higher output voltage for a given sound pressure input, and is suitable for use with vacuum-tube guitar amplifiers, for instance, which have a high input impedance and require a relatively high signal input voltage to overcome the tubes' inherent noise. Most professional microphones are low impedance, about 200 Ω or lower. Professional vacuum-tube sound equipment incorporates a transformer that steps up the impedance of the microphone circuit to the high impedance and voltage needed to drive the input tube. External matching transformers are also available that can be used in-line between a low impedance microphone and a high impedance input.

Low-impedance microphones are preferred over high impedance for two reasons: one is that using a high-impedance microphone with a long cable results in high frequency signal loss due to cable capacitance, which forms a low-pass filter with the microphone output impedance. The other is that long high-impedance cables tend to pick up more hum (and possibly radio-frequency interference (RFI) as well). Nothing is damaged if the impedance between microphone and other equipment is mismatched; the worst that happens is a reduction in signal or change in frequency response.

Some microphones are designed *not* to have their impedance matched by the load they are connected to. Doing so can alter their frequency response and cause distortion, especially at high sound pressure levels. Certain ribbon and dynamic microphones are exceptions, due to the designers' assumption of a certain load impedance being part of the internal electro-acoustical damping circuit of the microphone.

Digital Microphone Interface

Neumann D-01 digital microphone and Neumann DMI-8 8-channel USB Digital Microphone Interface

The AES42 standard, published by the Audio Engineering Society, defines a digital interface for microphones. Microphones conforming to this standard directly output a digital audio stream through an XLR or XLD male connector, rather than producing an analog output. Digital microphones may be used either with new equipment with appropriate input connections that conform to the AES42 standard, or else via a suitable interface box. Studio-quality microphones that operate in accordance with the AES42 standard are now available from a number of microphone manufacturers.

Measurements and Specifications

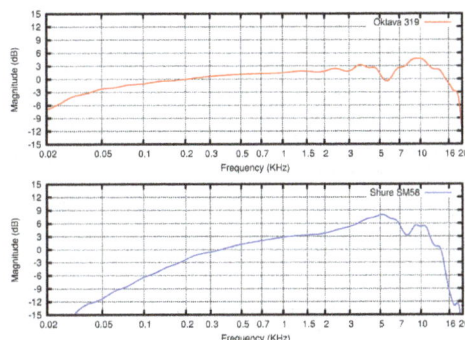

A comparison of the far field on-axis frequency response of the Oktava 319 and the Shure SM58

Because of differences in their construction, microphones have their own characteristic responses to sound. This difference in response produces non-uniform phase and frequency responses. In addition, microphones are not uniformly sensitive to sound pressure, and can accept differing levels without distorting. Although for scientific applications microphones with a more uniform response are desirable, this is often not the case for music recording, as the non-uniform response of a microphone can produce a desirable coloration of the sound. There is an international standard for microphone specifications, but few manufacturers adhere to it. As a result, comparison of published data from different manufacturers is difficult because different measurement techniques are used. The Microphone Data Website has collated the technical specifications complete with pictures, response curves and technical data from the microphone manufacturers for every currently listed microphone, and even a few obsolete models, and shows the data for them all in one common format for ease of comparison.. Caution should be used in drawing any solid conclusions from this or any other published data, however, unless it is known that the manufacturer has supplied specifications in accordance with IEC 60268-4.

A frequency response diagram plots the microphone sensitivity in decibels over a range of frequencies (typically 20 Hz to 20 kHz), generally for perfectly on-axis sound (sound arriving at 0° to the capsule). Frequency response may be less informatively stated textually like so: "30 Hz–16 kHz ±3 dB". This is interpreted as meaning a nearly flat, linear, plot between the stated frequencies, with variations in amplitude of no more than plus or minus 3 dB. However, one cannot determine from this information how *smooth* the variations are, nor in what parts of the spectrum they occur. Commonly made statements such as "20 Hz–20 kHz" are meaningless without a decibel measure of tolerance. Directional microphones' frequency response varies greatly with distance from the sound source, and with the geometry of the sound source. IEC 60268-4 specifies that frequency response should be measured in *plane progressive wave* conditions (very far away from the source) but this is seldom practical. *Close talking* microphones may be measured with different sound sources and distances, but there is no standard and therefore no way to compare data from different models unless the measurement technique is described.

The self-noise or equivalent input noise level is the sound level that creates the same output voltage as the microphone does in the absence of sound. This represents the lowest point of the microphone's dynamic range, and is particularly important should you wish to record sounds that are quiet. The measure is often stated in dB(A), which is the equivalent loudness of the noise on a decibel scale frequency-weighted for how the ear hears, for example: "15 dBA SPL" (SPL means sound pressure level relative to 20 micropascals). The lower the number the better. Some microphone manufacturers state the noise level using ITU-R 468 noise weighting, which more accurately represents the way we hear noise, but gives a figure some 11–14 dB higher. A quiet microphone typically measures 20 dBA SPL or 32 dB SPL 468-weighted. Very quiet microphones have existed for years for special applications, such the Brüel & Kjaer 4179, with a noise level around 0 dB SPL. Recently some microphones with low noise specifications have been introduced in the studio/entertainment market, such as models from Neumann and Røde that advertise noise levels between 5–7 dBA. Typically this is achieved by altering the frequency response of the capsule and electronics to result in lower noise within the A-weighting curve while broadband noise may be increased.

The maximum SPL the microphone can accept is measured for particular values of total harmonic distortion (THD), typically 0.5%. This amount of distortion is generally inaudible, so one can safely use the microphone at this SPL without harming the recording. Example: "142 dB SPL peak (at 0.5% THD)". The higher the value, the better, although microphones with a very high maximum SPL also have a higher self-noise.

The clipping level is an important indicator of maximum usable level, as the 1% THD figure usually quoted under max SPL is really a very mild level of distortion, quite inaudible especially on brief high peaks. Clipping is much more audible. For some microphones the clipping level may be much higher than the max SPL.

The dynamic range of a microphone is the difference in SPL between the noise floor and the maximum SPL. If stated on its own, for example "120 dB", it conveys significantly less information than having the self-noise and maximum SPL figures individually.

Sensitivity indicates how well the microphone converts acoustic pressure to output voltage. A high sensitivity microphone creates more voltage and so needs less amplification at the mixer or recording device. This is a practical concern but is not directly an indication of the microphone's quality, and in fact the term sensitivity is something of a misnomer, "transduction gain" being perhaps more meaningful, (or just "output level") because true sensitivity is generally set by the noise floor, and too much "sensitivity" in terms of output level compromises the clipping level. There are two common measures. The (preferred) international standard is made in millivolts per pascal at 1 kHz. A higher value indicates greater sensitivity. The older American method is referred to a 1 V/ Pa standard and measured in plain decibels, resulting in a negative value. Again, a higher value indicates greater sensitivity, so –60 dB is more sensitive than –70 dB.

Measurement Microphones

An AKG C214 condenser microphone with shock mount

Some microphones are intended for testing speakers, measuring noise levels and otherwise quantifying an acoustic experience. These are calibrated transducers and are usually supplied with a calibration certificate that states absolute sensitivity against frequency. The quality of measurement microphones is often referred to using the designations "Class 1," "Type 2" etc., which are references not to microphone specifications but to sound level meters. A more comprehensive standard for the description of measurement microphone performance was recently adopted.

Measurement microphones are generally scalar sensors of pressure; they exhibit an omnidirectional response, limited only by the scattering profile of their physical dimensions. Sound intensity or sound power measurements require pressure-gradient measurements, which are typically made using arrays of at least two microphones, or with hot-wire anemometers.

Calibration

To take a scientific measurement with a microphone, its precise sensitivity must be known (in volts per pascal). Since this may change over the lifetime of the device, it is necessary to regularly calibrate measurement microphones. This service is offered by some microphone manufacturers and by independent certified testing labs. All microphone calibration is ultimately traceable to primary standards at a national measurement institute such as NPL in the UK, PTB in Germany and NIST in the United States, which most commonly calibrate using the reciprocity primary standard. Measurement microphones calibrated using this method can then be used to calibrate other microphones using comparison calibration techniques.

Depending on the application, measurement microphones must be tested periodically (every year or several months, typically) and after any potentially damaging event, such as being dropped (most such microphones come in foam-padded cases to reduce this risk) or exposed to sounds beyond the acceptable level.

Arrays

A microphone array is any number of microphones operating in tandem. There are many applications:

- Systems for extracting voice input from ambient noise (notably telephones, speech recognition systems, hearing aids)

- Surround sound and related technologies

- Locating objects by sound: acoustic source localization, *e.g.*, military use to locate the source(s) of artillery fire. Aircraft location and tracking.

- High fidelity original recordings

- 3D spatial beamforming for localized acoustic detection of subcutaneous sounds

Typically, an array is made up of omnidirectional microphones distributed about the perimeter of a space, linked to a computer that records and interprets the results into a coherent form.

Windscreens

Windscreens (or windshields – the terms are interchangeable) provide a method of reducing the effect of wind on microphones. While pop-screens give protection from unidirectional blasts, foam "hats" shield wind into the grille from all directions, and blimps / zeppelins / baskets entirely enclose the microphone and protect its body as well. This last point is important because, given the extreme low frequency content of wind noise, vibration induced in the housing of the microphone can contribute substantially to the noise output.

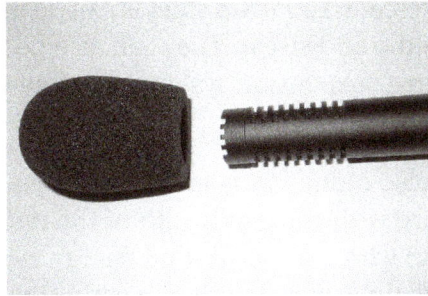

Microphone with its windscreen removed.

The shielding material used – wire gauze, fabric or foam – is designed to have a significant acoustic impedance. The relatively low particle-velocity air pressure changes that constitute sound waves can pass through with minimal attenuation, but higher particle-velocity wind is impeded to a far greater extent. Increasing the thickness of the material improves wind attenuation but also begins to compromise high frequency audio content. This limits the practical size of simple foam screens. While foams and wire meshes can be partly or wholly self-supporting, soft fabrics and gauzes require stretching on frames, or laminating with coarser structural elements.

Since all wind noise is generated at the first surface the air hits, the greater the spacing between shield periphery and microphone capsule, the greater the noise attenuation. For an approximately spherical shield, attenuation increases by (approximately) the cube of that distance. Thus larger shields are always much more efficient than smaller ones. With full basket windshields there is an additional pressure chamber effect, first explained by Joerg Wuttke, which, for two-port (pressure gradient) microphones, allows the shield/microphone combination to act as a high-pass acoustic filter.

Two recordings being made—a *blimp* is being used on the left. An open-cell foam windscreen is being used on the right.

Various microphone covers

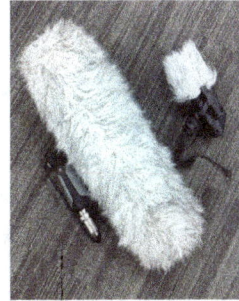

"Dead cat" and a "dead kitten" windscreens. The dead kitten covers a stereo microphone for a DSLR camera. The difference in name is due to the size of the fur.

Since turbulence at a surface is the source of wind noise, reducing gross turbulence can add to noise reduction. Both aerodynamically smooth surfaces, and ones that prevent powerful vortices being generated, have been used successfully. Historically, artificial fur has proved very useful for this purpose since the fibres produce micro-turbulence and absorb energy silently. If not matted by wind and rain, the fur fibres are very transparent acoustically, but the woven or knitted backing can give significant attenuation. As a material it suffers from being difficult to manufacture with consistency, and to keep in pristine condition on location. Thus there is an interest (DPA 5100, Rycote Cyclone) to move away from its use.

In the studio and on stage, pop-screens and foam shields can be useful for reasons of hygiene, and protecting microphones from spittle and sweat. They can also be useful coloured idents. On location the basket shield can contain a suspension system to isolate the microphone from shock and handling noise.

Stating the efficiency of wind noise reduction is an inexact science, since the effect varies enormously with frequency, and hence with the bandwidth of the microphone and audio channel. At very low frequencies (10–100 Hz) where massive wind energy exists, reductions are important to avoid overloading of the audio chain – particularly the early stages. This can produce the typical "wumping" sound associated with wind, which is often syllabic muting of the audio due to LF peak limiting. At higher frequencies – 200 Hz to ~3 kHz – the aural sensitivity curve allows us to hear the effect of wind as an addition to the normal noise floor, even though it has a far lower energy content. Simple shields may allow the wind noise to be 10 dB less apparent; better ones can achieve nearer to a 50 dB reduction. However the acoustic transparency, particularly at HF, should also be indicated, since a very high level of wind attenuation could be associated with very muffled audio.

Mixing Console

In audio, a mixing console is an electronic device for combining (also called "mixing"), routing, and changing the volume level, timbre (tone color) and/or dynamics of many different audio signals, such as microphones being used by singers, mics picking up acoustic instruments such as drums or saxophones, signals from electric or electronic instruments such as the electric bass or synthesizer, or recorded music playing on a CD player. In the 2010s, a mixer is able to control analog or digital signals, depending on the type of mixer. The modified signals (voltages or digital samples) are summed to produce the combined output signals, which can then be broadcast, amplified through a sound reinforcement system or recorded (or some combination of these applications).

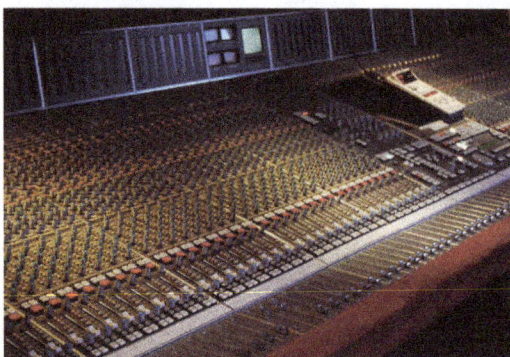

SSL SL9000J (72 channel) console at Cutting Room Recording Studio, NYC

An audio engineer adjusts a mixer while doing live sound for a band.

Mixing consoles are used in many applications, including recording studios, public address systems, sound reinforcement systems, nightclubs, dance clubs, broadcasting, television, and film post-production. A typical, simple application combines signals from two microphones (each used by vocalists singing a duet, perhaps) into an amplifier that drives one set of speakers simultaneously. In live performances, the signal from the mixer usually goes directly to an amplifier which is plugged into speaker cabinets, unless the mixer has a built-in power amplifier or is connected to powered speakers. A DJ mixer may have only two channels, for mixing two record players. A coffeehouse's tiny stage might only have a six channel mixer, enough for two singer-guitarists and a percussionist. A nightclub stage's mixer for rock music shows may have 24 channels for mixing the signals from a rhythm section, lead guitar and several vocalists. A mixing console for a large concert may have 48 channels. A mixing console in a professional recording studio may have as many as 96 channels.

In practice, mixers do more than simply mix signals. They can provide phantom power for capacitor microphones; pan control (which changes a sound's apparent position in the stereo soundfield); filtering and equalization, which enables sound engineers to boost or cut selected frequencies to improve the sound; routing facilities (to send the signal from the mixer to another device, such as a sound recording system or a control room; and monitoring facilities, whereby one of a number of sources can be routed to loudspeakers or headphones for listening, often without affecting the mixer's main output. Some mixers have onboard electronic effects, such as reverb. Some mixers intended for small venue live performance applications may include an integrated power amplifier.

Terminology

A mixing console is also known as an audio mixer, audio console, mixing desk, sound mixer, sound board, or simply as board or mixer.

Structure

Yamaha 2403 audio mixing console in a 'live' mixing application

A typical analog mixing board has three sections:

- Channel inputs
- Master controls
- Audio level metering

The channel input strips are usually a bank of identical monaural or stereo input channels. Each channel has rotary knobs, buttons and/or faders for controlling the gain and equalization (e.g., bass and treble) of the signal on each channel. Depending on the mixer, a channel may have buttons which enable the audio engineer to reroute the signal to a different output for monitoring purposes, turn on an attenuator "pad", or activate other features.

The master control section is used to adjust the levels of the overall output of the mixer. The master control section has sub-group faders, master faders, master auxiliary mixing bus level controls and auxiliary return level controls. In addition it may have solo monitoring controls, a stage "talk-back" microphone control (so the sound engineer can talk to the band, who may be some distance away at a live show or who might be separated in an isolation booth in the recording studio), muting controls and an output matrix mixer. On smaller mixers the inputs are on the left of the mixing board and the master controls are on the right. In larger mixers, the master controls are in the center with input faders and channel strips on both sides.

The audio level meters (which may be meters with needles or LEDs) may be above the input and master sections or they may be integrated into the input and master sections themselves. The audio level meters indicate when the signals are clipping.

Channel Input Strip

The input strip is usually separated into these sections:

- Input jacks

- Microphone preamplifiers

- equalization

- Dynamics processing (e.g. dynamic range compression, gating)

- Routing including direct outs, aux-sends, panning control and subgroup assignments

- Input faders (on some smaller mixers, these may be rotary knobs, to save space)

On many consoles, these sections are color-coded for quick identification by the operator. Each signal (e.g., a singer's vocal mic, the signal from an electric bass amp's DI box, etc.) that is plugged into the mixer has its own *channel*. Depending on the specific mixer, each channel is stereo or monaural. On most mixers, each channel has an XLR input, and many have RCA or quarter-inch TRS phone connector line inputs. The smallest, least expensive mixers may only have one XLR input with the other inputs being line inputs. These can be used by a singer-guitarist or other small acts.

Basic Input Controls

Below each input, there are usually several rotary controls (knobs or "pots"). The first knob is typically a *trim* or *gain* control. The inputs buffer the signal from the external device and this controls the amount of amplification (boosting) or attenuation (turning down of gain) needed to bring the signal to a nominal level for processing. This stage is where most noise of interference is picked up, due to the high gains involved (around +50 dB, for a microphone). Balanced inputs and connectors, such as XLR or phone connectors, reduce interference problems.

A mixing console may provide insert points after the buffer/gain stage. These provide a send and return connection for external processors that only affect an individual channel's signal. Effects that operate on multiple channels connect to auxiliary sends below.

Auxiliary Send Routing

The *auxiliary send* routes a split of the incoming signal to an auxiliary bus, which can then be routed to external devices. *Auxiliary sends* can either be pre-fader or post-fader, in that the level of a pre-fade send is set by the *auxiliary send* control, whereas post-fade sends depend on the position of the channel fader as well. *Auxiliary sends* can send the signal to an external processor such as a reverb, with the return signal routed through another channel or designated auxiliary return. These are normally post-fader. Pre-fade *auxiliary sends* can provide a monitor mix to musicians onstage (which they hear through monitor speakers pointing at the performers or in-ear monitors); this mix is thus independent of the main mix.

Programming channels on a radio sound board.

Most live radio broadcasting sound boards send audio through "program" channels. When a given channel button is selected, the audio will be sent to that device or transmitter. Program 1 is typically the on-air live feed, or what those listening to the broadcast will hear. Most boards have 3-4 programming channels, though some have more options. Often, one of the programming channels will feed one or more computers used for editing or sound playback. Another channel may be used to send audio to the talents' headset if they are broadcasting from a remote area.

Channel Equalization

Further channel controls affect the equalization of the signal by separately attenuating or boosting a range of frequencies. The smallest, least expensive mixers may only have bass and treble controls. Most mid-range and higher-priced mixers have bass, midrange, and treble, or even additional mid-range controls (e.g., low-mid and high-mid). Many high-end mixing consoles have a parametric equalizer on each channel. Some mixers have a general equalization control (either graphic or parametric) at the output, for controlling the tone of the overall mix.

Cue System

The cue system allows the operator to listen to one or more selected signals without affecting the console's audio outputs. A sound engineer can use the "cue" feature to get a sound recording she wishes to play soon cued up to the start point of a song, without the listeners hearing these actions. The signal from the cue system is fed to the console's headphone amp and may also be available as a line-level output that is intended to drive a monitor speaker system. The terms PFL (Pre Fade Listen) and AFL (After Fade Listen) are used to characterize the point in the signal flow from which the cue signal is derived. Input channels are usually configured as PFL so the operator can audition the channel without sending it to any mix. Consoles with a cue feature have a dedicated button on each channel, typically labeled *Cue* (or *AFL*, PFL, *Solo*, or *Listen*).

Solo In Place (SIP) is a related feature on advanced consoles. It typically is controlled by the Cue button, but unlike Cue, SIP is "destructive" (that is, has a major impact on) of the output mix. It mutes everything except channels being soloed. SIP is useful for setup and trouble-shooting, in that it allows the operator to quickly mute everything but the signal being worked on. SIP is potentially disastrous if engaged during performance, as it will mute all the channels except one, so most consoles require the operator to take very deliberate actions to engage SIP mode.

Subgroup and Mix Routing

Each channel on a mixer has a sliding volume control (*fader*) that allows adjustment of the level of that channel. Some smaller mixers may use a rotary control instead of a fader to save space. The signals are summed to create the main *mix*, or combined on a *bus* as a submix, a group of channels that are then added to get the final mix (for instance, many drum mics could be grouped into a bus, and then the proportion of drums in the final mix can be controlled with one bus fader). There may also be insert points for a certain bus, or even the entire mix.

VCA Groups

Some higher-end consoles use voltage-controlled amplifier (VCA) groups VCAs and DCAs function somewhat like subgroups but let the operator control the level of multiple input channels with a single fader. Unlike subgroups, no sub-mix is created. The audio signals from the assigned channels remain routed independently of VCA assignments. Since no sub-mix is created, it is not possible to insert processing such as compressors into a VCA/DCA group. In addition, on most VCA/DCA-equipped consoles, post-fader auxiliary send levels are affected by the VCA master. This is usually desirable, as post-fader auxiliary sends are commonly used for effects such as reverb, and sends to these effects should track changes in the channel signal level.

Master Output Controls

Subgroup and main output fader controls are often found together on the right hand side of the mixer or, on larger consoles, in a center section flanked by banks of input channels. Matrix routing is often contained in this master section, as are headphone and local loudspeaker monitoring controls. Talkback controls allow conversation with the artist through their monitors, headphones or in-ear monitor. A test tone generator might be located in the master output section. Aux returns such as those signals returning from external processors are often in the master section.

Metering

Meter bridge on *API Legacy Plus* console

Peak program meter   VU meter

Finally, there are usually one or more VU or peak meters (peak meters often use LEDs) to indicate the levels for each channel, for the master outputs and to indicate whether the console levels are clipping the signal. The sound engineer typically adjusts the gain of the input signals to get the strongest signal that can be obtained without causing "clipping" (unwanted distortion) or causing audio feedback "howls". Having the gain set as high as possible improves the signal to noise ratio. Most mixers have at least one additional output, besides the main mix. These are either individual bus outputs, or *auxiliary outputs*, used, for instance, to output a different mix to onstage monitors.

As the human ear experiences audio level in a logarithmic fashion (both amplitude and frequency), mixing console controls and displays are almost always in decibels, a logarithmic measurement system. Since it is a relative measurement, and not a unit itself, the meters must be referenced to a nominal level. The "professional" nominal level used on professional mixers is considered +4 dBu. The "consumer grade" level is –10 dBV.

Hardware Routing and Patching

For convenience, some mixing consoles include inserts or a patch bay or patch panel. Patch bays are mainly used for recording mixers. However, live sound mixers may also include patch bays. In live sound, the cables from the onstage microphones and instrument outputs are not typically plugged directly into the mixer, because this would require a large number of cables to go from the stage the mixer. Instead, the onstage mic and instrument cables are typically plugged into the patch bay of a thick snake cable, which runs from the stage to the mixer. The outputs from the snake's second patch bay (near the mixer) are then plugged into the mixer.

Other Features

Most, but not all, audio mixers can

- add external effects.

- use monaural signals to produce stereo sound through pan and balance controls.
- provide phantom power required by some microphones.

A sound engineer at the controls of a SSL9000J mixer.

Some mixers can

- create an audible "test tone" via an oscillator. The test tone can be used to troubleshoot issues before the band arrives and determine if channels are functioning properly.
- add effects internally.
- read and write console automation.
- be interfaced with computers or other recording equipment (to control the mixer with computer presets, for instance).
- control or be controlled by a digital audio workstation via MIDI or proprietary commands.
- be powered by batteries (this is only for the smallest mixers, such as four to six channel mixers that might be used on location outdoors).
- provide amplifier power for external speaker cabinets (these are called "powered mixers")

Mirroring

Some mixing consoles, particularly those designed for broadcast and live sound, include facilities for "mirroring" two consoles, making both consoles exact copies of each other with the same inputs and outputs, the same settings, and the same mix. There are two primary reasons for doing this; one, in the event of a hardware failure, a second redundant console is already in place and can be switched to (an important feature for live broadcasts); second, it allows the operators to set up two identical mix positions, one at front of house — where the audio will be mixed during a performance — and the other at some other location within the theater; this way, if the acoustics at front of house are unfavorable, a mix can be programmed at an acoustically better position in the room, and the presets can be accessed from the front of house console during the performance.

Digital versus Analog

Digital mixing console sales have increased dramatically since their introduction in the 1990s. Yamaha sold more than 1000 PM5D mixers by July, 2005, and other manufacturers are seeing increasing sales of their digital products. Digital mixers are more versatile than analog ones and offer many new features, such as reconfigure signal routing at the touch of a button. In addition, digital consoles often include processing capabilities such as compression, gating, reverb, automatic feedback suppression and delay. Some products are expandable via third-party software features (called plugins) that add further reverb, compression, delay and tone-shaping tools. Several digital mixers include spectrograph and real time analyzer functions. A few incorporate loudspeaker management tools such as crossover filtering and limiting. Digital signal processing can perform automatic mixing for some simple applications, such as courtrooms, conferences and panel discussions. Consoles with motorized faders can read and write console automation.

Digidesign's Venue Profile mixer on location at a corporate event. This digital mixer allows plugins from third-party vendors

Propagation Delay

Digital mixers have an unavoidable amount of latency or propagation delay, ranging from less than 1 ms to as much as 10 ms, depending on the model of digital mixer and what functions are engaged. This small amount of delay is not a problem for loudspeakers aimed at the audience or even monitor wedges aimed at the artist, but can be disorienting and unpleasant for IEMs (In-ear monitors) where the artist hears their voice acoustically in their head *and* electronically amplified in their ears but delayed by a couple of milliseconds.

Every analog to digital conversion and digital to analog conversion within a digital mixer entails propagation delay. Audio inserts to favorite external analog processors make for almost double the usual delay. Further delay can be traced to format conversions such as from ADAT to AES3 and from normal digital signal processing steps.

Within a digital mixer there can be differing amounts of latency, depending on the routing and on how much DSP is in use. Assigning a signal to two parallel paths with significantly different processing on each path can result in extreme comb filtering when recombined. Some digital mixers incorporate internal methods of latency correction so that such problems are avoided.

Ease of Use

In the 2010s, analog consoles remain popular, as they have a column of dedicated, physical knobs, buttons, and faders for each channel, which is logical and familiar to many users. This takes more physical space, but can accommodate rapid responses to changing performance conditions.

16-channel mixing console with compact short-throw faders

Most digital mixers use technology to reduce physical space requirements, entailing compromises in user interface such as a single shared channel adjustment area that is selectable for only one channel at a time. Additionally, most digital mixers have virtual pages or layers that change fader banks into separate controls for additional inputs or for adjusting equalization or aux send levels. This layering can be confusing for some operators. Analog consoles make for simpler understanding of hardware routing. Many digital mixers allow internal reassignment of inputs so that convenient groupings of inputs appear near each other at the fader bank, a feature that can be disorienting for persons having to make a hardware patch change.

On the other hand, many digital mixers allow for extremely easy building of a mix from saved data. USB flash drives and other storage methods are employed to bring past performance data to a new venue in highly portable manner. At the new venue, the traveling mix engineer simply plugs the collected data into the venue's digital mixer and quickly makes small adjustments to the local input and output patch layout, allowing for full show readiness in very short order. Some digital mixers allow offline editing of the mix, a feature that lets the traveling technician use a laptop to make anticipated changes to the show while *en route*, shortening the time it takes to prepare the sound system for the artist.

Sound Quality

A studio engineer at a Control 24 mixing surface.

Both digital and analog mixers rely on analog microphone preamplifiers, a high-gain circuit that increases the low signal level from a microphone to a level that is better matched to the console's internal operating level. In this respect, both formats are on par with each other. In a digital mixer, the microphone preamplifier is followed by an analog-to-digital converter. Ideally, this process is carefully engineered to deal gracefully with overloading and clipping while delivering an accurate digital stream. Further processing and mixing of digital streams within a mixer need to avoid saturation if maximum audio quality is desired.

Analog mixers, too, must deal gracefully with overloading and clipping at the microphone preamplifier and as well as avoiding overloading of mix buses. Very high frequency background hiss in an analog mixer is always present, though good gain stage management and turning unused channels down to zero minimizes its audibility. Idle subgroups left "up" in a mix add background hiss to the main outputs. Many digital mixers avoid this problem by low-level gating. Digital circuitry is more resistant to outside interference from radio transmitters such as walkie-talkies and cell phones. Hiss can be reduced with electronic noise reduction devices or with an equalizer.

Many electronic design elements combine to affect perceived sound quality, making the global "analog mixer vs. digital mixer" question difficult to answer. Experienced live sound professionals agree that the selection and quality of the microphones and loudspeakers (with their innate higher potential for creating distortion) are a much greater source of coloration of sound than the choice of mixer. The mixing style and experience of the person mixing may be more important than the make and model of audio console. Analog and digital mixers both have been associated with high-quality concert performances and studio recordings.

Remote Control

Hip hop producer Chilly Chill behind a large audio console in a recording studio.

Analog mixing in live sound has had the option since the 1990s of using wired remote controls for certain digital processes such as monitor wedge equalization and parameter changes in outboard reverb devices. That concept has expanded until wired and wireless remote controls are being seen in relation to entire digital mixing platforms. It is possible to set up a sound system and mix via laptop, touchscreen or tablet. Computer networks can connect digital system elements for expanded monitoring and control, allowing the system technician to make adjustments to distant devices during the performance. The use of remote control technology can be utilized to reduce "seat-kills", allowing more paying customers into the performance space.

Software Mixers

For recorded sound, the mixing process can be performed on screen, using computer software and associated input, output and recording hardware. The traditional large control surface of the mixing console is not utilized, saving space at the engineer's mix position. In a software studio, there is either no physical mixer fader bank at all or there is a compact group of motorized faders designed to fit into a small space and connected to the computer. Many project studios use such a space-efficient solution, as the mixing room at other times can serve as business office, media archival, etc. Software mixing is heavily integrated as part of a digital audio workstation.

Applications

A small four-channel mixer that could be used for a singer-guitarist's performance at a small coffeehouse.

Public address systems in schools, hospitals and other institutions use a mixing console to set microphones to an appropriate level and can add in recorded sounds such as music into the mix. PA mixers usually have controls that help to minimise audio feedback.

Most rock and pop bands use a mixing console to combine musical instruments and vocals so that the mix can be amplified through a nightclub's PA system. Among the highest quality bootleg recordings of live performances are so-called soundboard recordings sourced directly from the mixing console.

Radio broadcasts use a mixing desk to select audio from different sources, such as CD players, telephones, remote feeds, prerecorded advertisements, and in-studio live bands. These consoles, often referred to as "air-boards" are apt to have many fewer controls than mixers designed for live or studio production mixing, dropping pan/balance, EQ, and multi-bus monitoring/aux feed knobs in favor of cue and output bus selectors, since, in a radio studio, nearly all sources are either prerecorded or preadjusted.

DJs playing music for dancers at a dance club use a small DJ mixer to make smooth transitions between different songs which are played on sound sources that are plugged into the mixer. Compared with other mixers that are used in sound recording and live sound, DJ mixers have far fewer inputs. The most basic DJ mixers have only two inputs. Some DJ mixers have four or more inputs. These sound sources could be turntables, CD players, or iPods. The DJ mixer also allows the DJ to use headphones to cue the next song to the desired starting point before playing it.

Hip hop music DJs and Dub producers and engineers were early users of the mixing board as a musical instrument.

Noise music musicians may create feedback loops within mixers, creating an instrument known as a no-input mixer. The tones generated from a no-input mixer are created by connecting an output of the mixer into an input channel and manipulating the pitch with the mixer's dials.

Tape Recorder

A reel-to-reel tape recorder

An audio tape recorder, tape deck or tape machine is an analog audio storage device that records and plays back sounds, including articulated voices, usually using magnetic tape, either wound on a reel or in a cassette, for storage. In its present-day form, it records a fluctuating signal by moving the tape across a tape head that polarizes the magnetic domains in the tape in proportion to the audio signal. Tape-recording devices include reel-to-reel tape deck and the cassette deck.

The use of magnetic tape for sound recording originated around 1930. Magnetizable tape revolutionized both the radio broadcast and music recording industries. It gave artists and producers the power to record and re-record audio with minimal loss in quality as well as edit and rearrange recordings with ease. The alternative recording technologies of the era, transcription discs and wire recorders, could not provide anywhere near this level of quality and functionality. Since some early refinements improved the fidelity of the reproduced sound, magnetic tape has been the highest quality analog sound recording medium available. As of the first decade of the 21st century, analog magnetic tape has been largely replaced by digital recording technologies for consumer purposes, although some still record audio by analog.

Prior to the development of magnetic tape, magnetic wire recorders had successfully demonstrated the concept of magnetic recording, but they never offered audio quality comparable to the other recording and broadcast standards of the time. Some individuals and organizations developed innovative uses for magnetic wire recorders while others investigated variations of the technology. One particularly important variation was the application of an oxide powder to a long strip of paper. This German invention was the start of a long string of innovations that have led to present day magnetic tape recordings.

Wax Strip Recorder

The earliest known audio tape recorder was a non-magnetic, non-electric version invented by Alexander Graham Bell's Volta Laboratory and patented in 1886 (U.S. Patent 341,214). It employed a $\frac{3}{16}$-inch-wide (4.8 mm) strip of wax-covered paper that was coated by dipping it in a solution of beeswax and paraffin and then had one side scraped clean, with the other side allowed to harden. The machine was of sturdy wood and metal construction, and hand-powered by means of a knob fastened to the flywheel. The wax strip passed from one eight-inch reel around the periphery of a pulley (with guide flanges) mounted above the V-pulleys on the main vertical shaft, where it came in contact with either its recording or playback stylus. The tape was then taken up on the other reel. The sharp recording stylus, actuated by a vibrating mica diaphragm, cut the wax from the strip. In playback mode, a dull, loosely mounted stylus, attached to a rubber diaphragm, carried the reproduced sounds through an ear tube to its listener.

Both recording and playback heads, mounted alternately on the same two posts, could be adjusted vertically so that several recordings could be cut on the same $\frac{3}{16}$-inch-wide (4.8 mm) strip. While the machine was never developed commercially, it was an interesting ancestor to the modern magnetic tape recorder which it resembled somewhat in design. The tapes and machine created by Bell's associates, examined at one of the Smithsonian Institution's museums, became brittle, and the heavy paper reels warped. The machine's playback head was also missing. Otherwise, with some reconditioning, they could be placed into working condition.

The waxed tape recording medium was inferior to Edison's wax cylinder medium, and Edison's wax cylinder phonograph became the first widespread sound recording technology, used for both entertainment and office dictation.

Photoelectric Paper Tape Recorder

In 1932, after six years of developmental work, Merle Duston, a Detroit radio engineer, created a tape recorder that used a low-cost chemically treated paper tape, capable of recording both sounds and voice. During the recording process, the tape moved through a pair of electrodes which immediately imprinted the modulated sound signals as visible black stripes into the paper tape's surface. The sound track could be immediately replayed from the same recorder unit, which also contained photoelectric sensors, somewhat similar to the various motion picture sound-on-film technologies of the era.

On 13 August 1931, Duston filed USPTO Patent Application #556,743 for "Method Of And Apparatus For Electrically Recording And Reproducing Sound And Other Vibrations", and which was renewed in 1934.

Magnetic Recording

Magnetic recording was conceived as early as 1878 by the American engineer Oberlin Smith and demonstrated in practice in 1898 by Danish engineer Valdemar Poulsen. Analog magnetic wire recording, and its successor, magnetic tape recording, involve the use of a magnetizable medium which moves with a constant speed past a recording head. An electrical signal, which is analogous to the sound that is to be recorded, is fed to the recording head, inducing a pattern of magnetization similar to the signal. A playback head can then pick up the changes in magnetic field from the tape and convert it into an electrical signal to be amplified and played back through a loudspeaker.

Wire Recorders

Magnetic wire recorder, invented by Valdemar Poulsen, 1898. It is exhibited at Brede works
Industrial Museum, Lyngby, Denmark.

The first wire recorder was the Telegraphone invented by Valdemar Poulsen in the late 1890s. Wire recorders for law/office dictation and telephone recording were made almost continuously by various companies (mainly the American Telegraphone Company) through the 1920s and 1930s. These devices were mostly sold as consumer technologies after World War II.

Widespread use of the wire recording device occurred within the decades spanning from 1940 until 1960, following the development of inexpensive designs licensed internationally by the Brush Development Company of Cleveland, Ohio and the Armour Research Foundation of the Armour Institute of Technology (later Illinois Institute of Technology). These two organizations licensed dozens of manufacturers in the U.S., Japan, and Europe. Wire was also used as a recording medium in black box voice recorders for aviation in the 1950s.

Consumer wire recorders were marketed for home entertainment or as an inexpensive substitute for commercial office dictation recorders, but the development of consumer magnetic tape recorders starting in 1946, with the BK 401 Soundmirror, using paper-based tape, quickly drove wire recorders from the market.

Early Steel Tape Recorders

Blattnerphone steel tape recorder at BBC studios, London, 1937

In 1924 a German engineer, Dr. Kurt Stille, developed the Poulsen wire recorder as a dictating machine. The following year a fellow German, Louis Blattner, working in Britain, licensed Stille's device and started work on a machine which would instead record on a magnetic steel tape, which he called the Blattnerphone.

The BBC installed a Blattnerphone at Avenue House in September 1930 for tests, and used it to record King George V's speech at the opening of the India Round Table Conference on 12 November 1930. Though not considered suitable for music the machine continued in use and was moved to Broadcasting House in March 1932, a second machine also being installed.

The tape was 6mm wide and 0.08mm thick, travelling at 5 feet per second; the recording time was 20 minutes.

In September 1932 a new model was installed, using 3mm tape with a recording time of 32 minutes.

In 1933 the Marconi Company purchased the rights to the Blattnerphone, and newly developed Marconi-Stille recorders were installed in the BBC's Maida Vale Studios in March 1935. The quality was slightly improved, though it still tended to be obvious that one was listening to a recording, as was the reliability. A reservoir system containing a loop of tape helped to stabilize the speed (there was also a smaller one just before the heads). The tape was 3mm wide and travelled at 1.5 metres/second. By September there were three recording rooms, each with two machines.

They were hardly easy to handle. The spools were heavy (and expensive) and the tape has been described as being like a travelling razor blade. The tape was liable to snap, particularly at joins, which at that speed could rapidly cover the floor with loops of the sharp-edged tape. Rewinding was done at twice the speed of the recording.

However, despite all this, the ability to make replayable recordings was extremely useful, and even with subsequent methods coming into use (direct-cut discs and Philips-Miller optical film) the Marconi-Stilles remained in use until the late 1940s.

Modern Tape Recorders

Magnetophon from a German radio station in World War II..

Magnetic tape recording as we know it today was developed in Germany during the 1930s at BASF (then part of the chemical giant IG Farben) and AEG in cooperation with the state radio RRG. This was based on Fritz Pfleumer's 1928 invention of paper tape with oxide powder lacquered to it. The first practical tape recorder from AEG was the Magnetophon K1, demonstrated in Germany in 1935. Eduard Schüller of AEG built the recorders and developed a ring shaped recording and playback head. It replaced the needle shaped head which tended to shred the tape. Friedrich Matthias of IG Farben/BASF developed the recording tape, including the oxide, the binder, and the backing material. Walter Weber, working for Hans Joachim von Braunmühl at the RRG, discovered the AC biasing technique, which radically improved sound quality.

During World War II, the Allies noticed that certain German officials were making radio broadcasts from multiple time zones almost simultaneously. Analysts such as Richard H. Ranger believed that the broadcasts had to be transcriptions, but their audio quality was indistinguishable from that of a live broadcast and their duration was far longer than was possible even with 16 rpm transcription discs. (The Allies were aware of the existence of the pre-war Magnetophon recorders, but not of the introduction of high-frequency bias and PVC-backed tape.) In the final stages of the war in Europe, the Allied capture of a number of German Magnetophon recorders from Radio Luxembourg aroused great interest. These recorders incorporated all the key technological features of modern analog magnetic recording and were the basis for future developments in the field.

Commercialization

American Developments

Development of magnetic tape recorders in the late 1940s and early 1950s is associated with the Brush Development Company and its licensee, Ampex. The equally important development of the magnetic tape media itself was led by Minnesota Mining and Manufacturing (3M) corporation.

In 1938, S.J. Begun left Germany and joined the Brush Development Company in the United States, where work continued but attracted little attention until the late 1940s when the company released the very first consumer tape recorder in 1946: the Soundmirror BK 401. Several other models were quickly released in the following years. Tapes were initially made of paper coated with magnetite powder. Minnesota Mining & Manufacturing Company (3M) replaced them by plastic tapes in 1948.

American audio engineer John T. Mullin and entertainer Bing Crosby were key players in the commercial development of magnetic tape. Mullin served in the U.S. Army Signal Corps and was posted to Paris in the final months of WWII. His unit was assigned to find out everything they could about German radio and electronics, including the investigation of claims that the Germans had been experimenting with high-energy directed radio beams as a means of disabling the electrical systems of aircraft. Mullin's unit soon amassed a collection of hundreds of low-quality magnetic dictating machines, but it was a chance visit to a studio at Bad Nauheim near Frankfurt while investigating radio beam rumours, that yielded the real prize.

Mullin was given two suitcase-sized AEG 'Magnetophon' high-fidelity recorders and fifty reels of recording tape. He had them shipped home and over the next two years he worked on the machines constantly, modifying them and improving their performance. His major aim was to interest Hollywood studios in using magnetic tape for movie soundtrack recording.

Mullin gave two public demonstrations of his machines, and they caused a sensation among American audio professionals; many listeners literally could not believe that what they heard was not a live performance. By luck, Mullin's second demonstration was held at MGM studios in Hollywood and in the audience that day was Bing Crosby's technical director, Murdo Mackenzie. He arranged for Mullin to meet Crosby and in June 1947 he gave Crosby a private demonstration of his magnetic tape recorders.

Bing Crosby's Influence

Bing Crosby, a top movie and singing star, was stunned by the amazing sound quality and instantly saw the huge commercial potential of the new machines. Live music was the standard for American radio at the time and the major radio networks didn't permit the use of disc recording in many programs because of their comparatively poor sound quality. Crosby disliked the regimentation of live broadcasts 39 weeks a year, preferring the recording studio's relaxed atmosphere and ability to retain the best parts of a performance. He had asked NBC to let him pre-record his 1944-45 series on transcription discs, but the network refused, so Crosby had withdrawn from live radio for a year. ABC agreed to let him use transcription discs for the 1946-47 season, but listeners complained about the sound quality.

Mullin's tape recorder came along at precisely the right moment. Crosby realised that the new technology would enable him to pre-record his radio show with a sound quality that equalled live broadcasts, and that these tapes could be replayed many times with no appreciable loss of quality. Mullin was asked to tape one show as a test and was immediately hired as Crosby's chief engineer to pre-record the rest of the series.

Crosby's season premier on 1 October 1947 was the first magnetic tape broadcast in America. He became the first major American music star to use tape to pre-record radio broadcasts, and the first to master commercial recordings on tape. The taped Crosby radio shows were painstakingly edited through tape-splicing to give them a pace and flow that was wholly unprecedented in radio. Mullin even claims to have been the first to use "canned laughter"; at the insistence of Crosby's head writer, Bill Morrow, he inserted a segment of raucous laughter from an earlier show into a joke in a later show that hadn't worked well. Soon other radio performers were demanding the ability to prerecord their broadcasts with the high quality of tape, and the recording ban was lifted.

Keen to make use of the new recorders as soon as possible, Crosby invested $50,000 of his own money into the Californian electronics company Ampex, and the tiny six-man concern (headed by Alexander M. Poniatoff, whose initials became part of the company name) soon became the world leader in the development of tape recording, revolutionising radio and recording with its famous Model 200 tape deck, issued in 1948 and developed directly from Mullin's modified Magnetophons.

Tape Recording at the BBC

The BBC acquired some Magnetophon machines in 1946 on an experimental basis, and these were used in the early stages of the new Third Programme to record and play back performances of operas from Germany (live relays being problematic because of the unreliability of the landlines in the immediate post-war period).

These machines were used until 1952, though most of the work continued to be done using the established media; but from 1948 a new British model became available from EMI: The BTR1.

Though in many ways clumsy, its quality was good, and as it wasn't possible to obtain any more Magnetophons it was an obvious choice.

EMI BTR2 machines in a BBC recording room, 12 November 1961.

In 1963 The Beatles were allowed to enhance their recordings at the BBC by overdubbing. The BBC didn't have any multi track tapes. They would copy them onto another tape.

In the early 1950s the EMI BTR 2 became available (right); a much improved machine and generally liked. It became the standard in recording channels (rooms) for many years, and was in use until the end of the 1960s.

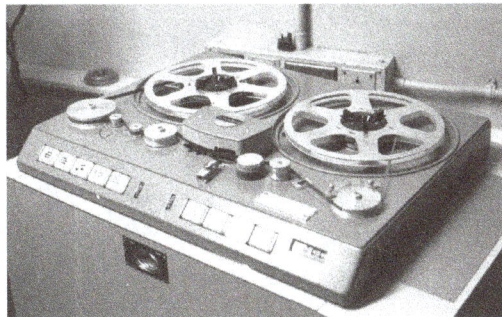

Early model Studer professional tape recorder, 1969

The machines were responsive, could run up to speed quite quickly, had light-touch operating buttons, forward-facing heads (The BTR 1s had rear-facing heads which made editing difficult), and were quick and easy to do the finest editing on.

The tape speed was eventually standardized at 15 ips for almost all work at Broadcasting House, and at 15 ips for music and 7½ ips for speech at Bush House. Broadcasting House also used the EMI TR90 and a Philips machine which was lightweight but very easy and quick to use: Bush House used several Leevers-Rich models.

The Studer range of machines had become pretty well the studio recording industry standard by the 1970s, and gradually these replaced the ageing BTR2s in recording rooms and studios. By the mid-2000s tape was pretty well out of use and had been replaced by digital playout systems.

Standardized Products

Working with Mullin in The USA, Ampex rapidly developed two-track stereo and then three-track recorders.

The typical professional audio tape recorder of the early 1950s used $\frac{1}{4}$ in (6 mm) wide tape on 10 $\frac{1}{2}$ in (27 cm) reels, with a capacity of 2,400 ft (730 m). Typical speeds were initially 15 in/s (38.1 cm/s) yielding 30 minutes› recording time on a 2,400 ft (730 m) reel. 30 in/s (76.2 cm/s) was used for the highest quality work. Domestic and portable recorders used 7, 5 or 3 in (18, 13 or 8 cm) inch reels (or spools). Early professional machines used single-sided spools but double-sided spools soon became popular (particularly for domestic use). Tape spools were usually made from transparent plastic but metal spools were also used.

Standard tape speeds varied by factors of two — 15 and 30 in/s were used for professional audio recording; 7 $\frac{1}{2}$ in/s (19.1 cm/s) for home audiophile prerecorded tapes; 7 $\frac{1}{2}$ and 3 $\frac{3}{4}$ in/s (19.1 and 9.5 cm/s) for audiophile and consumer recordings (typically on 7 in (18 cm) reels). 1 $\frac{7}{8}$ in/s (4.8 cm/s) and occasionally even $\frac{15}{16}$ in/s (2.4 cm/s) were used for voice, dictation, and applications where very long recording times were needed, such as logging police and fire department calls.

The 8-track tape standard, promoted by Bill Lear in the early 1960s, popularized consumer audio playback in automobiles. Eventually, this standard was replaced by the smaller and more reliable Compact Cassette.

Compact cassette

Philips' development of the Compact Cassette in 1963 and Sony's development of the Walkman in 1979 led to widespread consumer use of magnetic audio tape. In 1990, the Compact Cassette was the dominant format in mass-market recorded music. The development of Dolby noise reduction technology in the 1960s brought audiophile quality recording to the Compact Cassette also contributing to its popularity.

Later Developments

Since their first introduction, analog tape recorders have experienced a long series of progressive developments resulting in increased sound quality, convenience, and versatility.

- Two-track and, later, multi-track heads permitted discrete recording and playback of individual sound sources, such as two stereophonic channels, or different microphones during live recording. The more versatile machines could be switched to record on some tracks while playing back others, permitting additional tracks to be "laid down" in synchronisation with previously recorded material such as a rhythm track.

- Use of separate heads for recording vs. playback (three heads total, counting the erase head) enabled monitoring of the recorded signal a fraction of a second after recording. Mixing the playback signal back into the record input also created a primitive echo generator.

- Dynamic range compression during recording and expansion during playback expanded the available dynamic range and improved the signal-to-noise ratio. dbx and Dolby Laboratories introduced add-on products in this area, originally for studio use, and later in versions for the consumer market. In particular, "Dolby B" noise reduction became very common in all but the least expensive cassette tape recorders.

Solidyne GMS200 tape recorder with computer self-adjustment. Argentina 1980–1990

- Computer-controlled analog tape recorders were introduced by Oscar Bonello in Argentina. The mechanical transport used three DC motors and introduced two new advances: automated microprocessor transport control and automatic adjustment of bias and frequency response. In 30 seconds the recorder adjusted its bias for minimum THD and best frequency response to match the brand and batch of magnetic tape used. The microprocessor control of transport allowed fast location of any point on the tape.

Operation

Electrical

Electric current flowing in the coils of the tape head creates a fluctuating magnetic field. This causes the magnetic material on the tape, which is moving past and in contact with the head, to align in a manner proportional to the original signal. The signal can be reproduced by running the tape back across the tape head, where the reverse process occurs – the magnetic imprint on the tape induces a small current in the read head which approximates the original signal and is then amplified for playback. Many tape recorders are capable of recording and playing back at once by means of separate record and playback heads in line or combined in one unit.

Mechanical

Modern professional recorders usually use a three-motor scheme. One motor with a constant rotational speed drives the capstan. This, usually combined with a rubber pinch roller, ensures that the tape speed does not fluctuate. The other two motors, which are called Torque Motors, apply equal and opposite torques to the supply and take up reels during recording and play back functions and maintain the tape's tension. During fast winding operations the pinch roller is disengaged and the take up reel motor is supplied with a higher voltage than the supply motor. The cheapest models use a single motor for all required functions; the motor drives the capstan directly and the supply and take-up reels are loosely coupled to the capstan motor with slipping belts or clutches. There are also variants with two motors, in which one motor is used for rewinding only.

A typical portable desktop cassette recorder from RadioShack.

Limitations

The storage of an analogue signal on tape works well, but is not perfect. In particular, the granular nature of the magnetic material adds high-frequency noise to the signal, generally referred to as tape hiss. Also, the magnetic characteristics of tape are not linear. They exhibit a characteristic hysteresis curve, which causes unwanted distortion of the signal. Some of this distortion is overcome by using an inaudible high-frequency AC bias signal when recording, though the amount of bias needs careful adjustment for best results. Different tape material requires differing amounts of bias, which is why most recorders have a switch to select this (or, in a cassette recorder, switch automatically based on cutouts in the cassette shell). Additionally, systems such as Dolby noise reduction systems (Dolby B, Dolby C, Dolby S and Dolby HX-Pro) have been devised to ameliorate some noise and distortion problems. Variations in tape speed cause flutter, which can be reduced by using dual capstans. Higher speeds used in professional recorders are prone to cause "head bumps," which are fluctuations in low-frequency response.

Tape Recorder Variety

Otari MX-80 24-track with 2-inch reels

There are a wide variety of tape recorders in existence, from small hand-held devices to large multitrack machines. A machine with built-in speakers and audio power amplification to drive them is usually called a "tape recorder" or – if it has no record functionality – a "tape player," while one that requires external amplification for playback is usually called a "tape deck" (regardless of whether it can record).

Multitrack technology enabled the development of modern art music and one such artist, Brian Eno, described the tape recorder as "an automatic musical collage device".

Uses

Magnetic tape brought about sweeping changes in both radio and the recording industry. Sound could be recorded, erased and re-recorded on the same tape many times, sounds could be duplicated from tape to tape with only minor loss of quality, and recordings could now be very precisely edited by physically cutting the tape and rejoining it. In August 1948, Los Angeles-based Capitol Records became the first recording company to use the new process.

Klaudia Wilde from the German WDR radio archive with a broadcast tape from 1990.
This is a centre hub with only a very short length of tape wound round it.

Within a few years of the introduction of the first commercial tape recorder, the Ampex 200 model, launched in 1948, American musician-inventor Les Paul had invented the first multitrack tape recorder, bringing about another technical revolution in the recording industry. Tape made possible the first sound recordings totally created by electronic means, opening the way for the bold sonic experiments of the Musique Concrète school and avant garde composers like Karlheinz Stockhausen, which in turn led to the innovative pop music studio recordings of artists such as Frank Zappa, The Beatles and The Beach Boys.

Editing magnetic tape

Tape enabled the radio industry for the first time to pre-record many sections of program content such as advertising, which formerly had to be presented live, and it also enabled the creation and duplication of complex, high-fidelity, long-duration recordings of entire programs. It also, for the first time, allowed broadcasters, regulators and other interested parties to undertake comprehensive logging of radio broadcasts for legislative and commercial purposes, leading to the growth of the modern media monitoring industry.

Innovations, like multitrack recording and tape echo, enabled radio programs and advertisements to be pre-produced to a level of complexity and sophistication that was previously unattainable and tape also led to significant changes to the pacing of program content, thanks to the introduction of the endless-loop tape cartridge.

An important use of tape recorders is the recording of video. Video cassette recorders differ substantially from audio recorders due to the use of a rotating magnetic head that uses a helical scan over the tape medium. Helical scans increase the relative speed of the tape surface over the head.

While they are primarily used for sound recording, tape machines were also important for data storage before the advent of floppy disks and CDs, and are still used today, although primarily to provide an offline backup to hard disk drives.

Tapedeck Speeds

Professional decks will use higher tape speeds, with 15 and 30 inches per second being most common, while lower tape speeds are usually used for smaller recorders and cassette players, in order to save space where fidelity is not as critical as in professional recorders. By providing a range of tape speeds, users can trade-off recording time against signal quality with higher tape speeds providing greater frequency response.

There are many tape speeds which are in use in all sorts of tape recorders. Speed may be expressed in centimeters per second (cm/s) or in inches per second (in/s).

Common tape speeds	
cm/s	in/s
1.2	15/32
2.4	15/16
4.75	$1\frac{7}{8}$
9.5	$3\frac{3}{4}$
19	$7\frac{1}{2}$
38	15
76	30

Music Sequencer

A music sequencer (or simply sequencer) is a device or application software that can record, edit, or play back music, by handling note and performance information in several forms, typically CV/

Gate, MIDI, or Open Sound Control (OSC), and possibly audio and automation data for DAWs and plug-ins.

Overview

Modern Sequencers

The advent of Musical Instrument Digital Interface (MIDI) and the Atari ST home computer in the 1980s gave programmers the opportunity to design software that could more easily record and play back sequences of notes played or programmed by a musician. This software also improved on the quality of the earlier sequencers which tended to be mechanical sounding and were only able to play back notes of exactly equal duration. Software-based sequencers allowed musicians to program performances that were more expressive and more human. These new sequencers could also be used to control external synthesizers, especially rackmounted sound modules, and it was no longer necessary for each synthesizer to have its own devoted keyboard.

1980s typical software sequencer platform, using Atari Mega ST computer.

Today's typical software sequencer, supporting multitrack audio (DAW) and plug-ins (Steinberg Cubase 6

As the technology matured, sequencers gained more features, such as the ability to record multi-track audio. Sequencers used for audio recording are called digital audio workstations (or DAWs).

Many modern sequencers can be used to control virtual instruments implemented as software plug-ins. This allows musicians to replace expensive and cumbersome standalone synthesizers with their software equivalents.

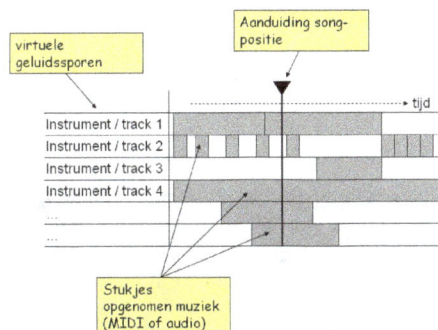

User interface on Steinberg Cubase v6.0, a digital audio workstation with an integrated software sequencer.

Today the term "sequencer" is often used to describe software. However, hardware sequencers still exist. Workstation keyboards have their own proprietary built-in MIDI sequencers. Drum machines and some older synthesizers have their own step sequencer built in. There are still also standalone hardware MIDI sequencers, although the market demand for those has diminished greatly due to the greater feature set of their software counterparts.

Types of Music Sequencer

Music sequencers are often categorized by handling data types, as following:.

- MIDI data on the MIDI sequencers (implemented as hardware or software)

- CV/Gate data on the analog sequencers and possibly others (via CV/Gate interfaces)

- Automation data for mixing-automation on the DAWs, and the software effect / instrument plug-ins on the DAWs with sequencing features

- Audio data on the audio sequencers including DAW, loop-based music software, etc.; or, the phrase samplers including Groove machines, etc.

Also, music sequencer can be categorized by its construction and supporting modes.

Realtime Sequencer (Realtime Recording Mode)

A realtime sequencer on the synthesizer

Realtime sequencers record the musical notes in real-time as on audio recorders, and play back musical notes with designated tempo, quantizations, and pitch. For editing, usually "punch in/ punch out" feature originated in the tape recording is provided, although it requires enough skills to obtain desired result. For detailed editing, possibly another visual editing modes under graphical user interface may be more suitable. Anyway, this mode provides usability similar to the audio recorder already familiarized by musicians, and it is widely supported on software sequencer, DAW, and built-in hardware sequencers.

Analog Sequencer

An analog sequencer

Analog sequencers are typically implemented with analog electronics, and play the musical notes designated by a series of knobs or sliders corresponding to each musical note (step). It is designed for both composition and live performance; users can change the musical notes at any time without regarding recording mode. And also possibly, the time-interval between each musical note (length of each step) can be independently adjustable. Typically, analog sequencer is used to generate the repeated minimalistic phrases which is reminiscent of Tangerine Dream, Giorgio Moroder or trance music.

Step Sequencer (Step Recording Mode)

A step rhythm sequencer on the drum machine | A step note sequencer on the bass machine

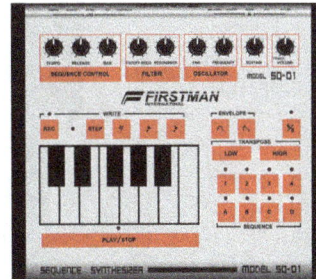

On the step sequencers, musical notes are rounded into the steps of equal time-interval, and users can enter each musical note without exact timing; Instead, each timing and duration of step are designated in several ways:

- On the drum machines: select a trigger timing from a row of step-buttons.

- On the bass machines: select a step note (or rest) from a chromatic keypads, then select a step duration (or tie) from a group of length-buttons, sequentially.

- On the several home keyboards: in addition to the realtime sequencer, a pair of step trigger button is provided; using it, notes on the pre-recorded sequence can be triggered in arbitrary timings for the timing dedicated recordings or performances.

In general, step mode, along with roughly quantized semi-realtime mode, is often supported on the analog drum machines, bass machines and several groove machines.

Software Sequencer

Software sequencer is a class of application software providing a functionality of music sequencer, and often provided as one feature of the DAW or the integrated music authoring environments. The features provided as sequencers vary widely depending on the software; even an analog sequencer can be simulated. The user may control the software sequencer either by using the graphical user interfaces or a specialized input devices, such as a MIDI controller.

History

Early Sequencers

The early music sequencers were sound producing devices such as automatic musical instruments, music boxes, mechanical organs, player pianos, and Orchestrions. Player pianos, for example, had much in common with contemporary sequencers. Composers or arrangers transmitted music to piano rolls which were subsequently edited by technicians who prepared the rolls for mass duplication. Eventually consumers were able to purchase these rolls and play them back on their own player pianos.

Barrel with pins on a large stationary barrel organ

The origin of automatic musical instruments seems considerably old. As early as the 9th century, Persian inventors Banū Mūsā brothers invented hydropowered organ using exchangeable cylinders with pins, and also automatic flute player using steam power, as described on their *Book of Ingenious Devices*. In the 14th century, rotating cylinder with pins were used to play carillon in Flanders, and at least in the 15th century, barrel organs were seen in the Netherlands.

Music roll on barrel organ

In the late-18th or early-19th century, as the results of Industrial Revolution, various automatic musical instruments were invented, for examples: music box, barrel organ and barrel piano using barrel / cylinder with pins or metal disc with punched holes; or mechanical organ, player piano and orchestrion using book music / music rolls (piano rolls) with punched holes, etc. These instruments were widely spread as the popular entertainment devices before the inventions of phonograph, radio, and sound film. Amongst of all, especially the punched tape media had been long lived until the mid-20th century: earliest programmable music synthesizers including RCA Mark II Sound Synthesizer in 1957, and Siemens Synthesizer in 1959, were also controlled via punch tapes similar to piano rolls.

Another inventions were came from sound film technology. The drawn sound technique which appeared in the late 1920s, is notable as a precursor of today's intuitive graphical user interfaces. On this technique, notes and various sound parameters were controlled by hand-drawn waves on the films, resembling piano rolls or strip charts on the modern sequencers/DAWs. It was often utilized on early experiments of electronic music, including Variophone developed by Yevgeny Sholpo in 1930, and Oramics designed by Daphne Oram in 1957, etc.

Analog Sequencers

Earliest commercially available analog sequencers (bottom) on Buchla 100 (1964/1966)
Moog sequencer module (left, probably added after 1968) on Moog Modular (1964)

During the 1940s–1960s, Raymond Scott, an American composer of electronic music, invented various kind of music sequencers for his electric compositions. The "Wall of Sound", once covered on the wall of his studio in New York during the 1940s–1950s, was an electro-mechanical sequencer to produce rhythmic patterns, consisting of stepping relays (used on dial pulse telephone exchange), solenoids, control switches, and tone circuits with 16 individual oscillators. Later, Robert Moog explained it as "the whole room would go 'clack - clack - clack', and the sounds would come out all over the place". The Circle Machine, developed in 1959, had dimmer bulbs arranged in a ring, and a rotating arm with photocell scanning over the ring, to generate arbitrary waveform. Also, the rotating speed of arm was controlled via brightness of lights, and as the results, arbitrary rhythms were generated.

Clavivox, developed since 1952, was a kind of keyboard synthesizer with sequencer.On its prototype, a theremin manufactured by young Robert Moog was utilized to enable portamento over 3-octave range, and on later version, it was replaced by a pair of photographic film and photocell for controlling the pitch by voltage.

In 1965 Ralph Lundsten had a polyphonic synthesizer with sequencer called Andromatic. built for him by Erkki Kurenniemi.

Step Sequencers

Electro-mechanical disc sequencer on early drum machine (1959)

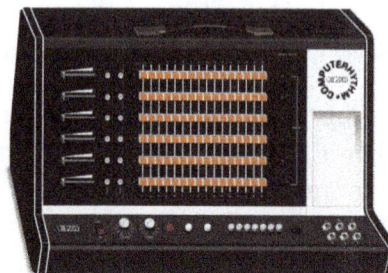

Eko ComputeRhythm (1972), one of the earliest programmable drum machines

The step sequencers played rigid patterns of notes using a grid of (usually) 16 buttons, or steps, each step being 1/16 of a measure. These patterns of notes were then chained together to form longer compositions. Sequencers of this kind are still in use, mostly built into drum machines and grooveboxes. They are monophonic by nature, although some are multi-timbral, meaning that they can control several different sounds but only play one note on each of those sounds.

Early Computers

On the other hand, software sequencers were continuously utilized since the 1950s in the context of computer music, including computer-*played* music (software sequencer), computer-*composed* music (music synthesis), and computer *sound generation* (sound synthesis). In June 1951, the first computer music *Colonel Bogey* was played on CSIRAC, Australia's first digital computer. In 1956, Lejaren Hiller at the University of Illinois at Urbana-Champaign wrote one of the earliest programs for computer music composition on ILLIAC, and collaborated on the first piece, *Illiac Suite for String Quartet*, with Leonard Issaction. In 1957 Max Mathews at Bell Labs wrote MUSIC, the first widely used program for sound generation, and a 17-second composition was performed by the IBM 704 computer. Subsequently, computer music was mainly researched on the expensive mainframe computers in computer centers, until the 1970s when minicomputers and then microcomputers became available in this field.

CSIRAC played the earliest computer music in 1951

In Japan, experiments in computer music date back to 1962, when Keio University professor Sekine and Toshiba engineer Hayashi experimented with the TOSBAC computer. This resulted in a piece entitled *TOSBAC Suite*.

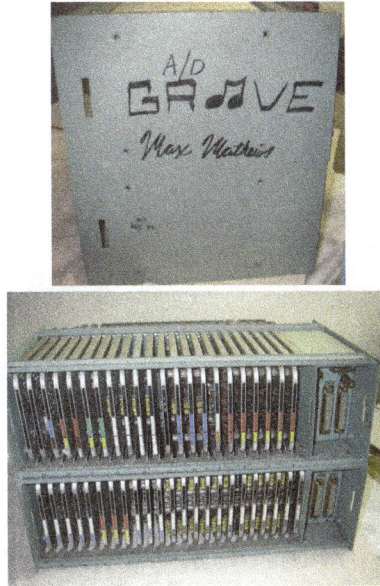

DDP-24 S Block (expansion card rack unit) that is assumed the A/D converters used
for GROOVE (1970) by Max Mathews.

In 1965, Mathews and L. Rosler developed Graphic 1, an interactive graphical sound system (that implies sequencer) on which one could draw figures using a light-pen that would be converted into sound, simplifying the process of composing computer generated music. It used PDP-5 mini-computer for data input, and IBM 7094 mainframe computer for rendering sound. Also in 1970, Mathews and F. R. Moore developed the GROOVE (Generated Real-time Output Operations on Voltage-controlled Equipment) system, a first fully developed music synthesis system for interactive composition (that implies sequencer) and realtime performance, using 3C/Honeywell DDP-24 (or DDP-224) minicomputers. It used a CRT display to simplify the management of music synthesis in realtime, 12bit D/A for realtime sound playback, an interface for analog devices, and even several controllers including a musical keyboard, knobs, and rotating joysticks to capture realtime performance.

Digital Sequencers

In 1971, Electronic Music Studios (EMS) released one of the first digital sequencer products as a module of Synthi 100, and its derivation, Synthi Sequencer series. After then, Oberheim released the DS-2 Digital Sequencer in 1974, and Sequential Circuits released Model 800 in 1977.

EMS Sequencer 256 (1971), branched from Synthi 100.

Music Workstations

In 1975, New England Digital (NED) released ABLE computer (microcomputer) as a dedicated data processing unit for Dartmouth Digital Synthesizer (1973), and based on it, later Synclavier series were developed.

Synclavier I (1977)

Fairlight CMI (1979) supporting MCL (sequencer)

The Synclavier I, released in September 1977, was one of the earliest digital music workstation product with multitrack sequencer. Synclavier series evolved throughout the late-1970s to the mid-1980s, and they also established integration of digital-audio and music-sequencer, on their Direct-to-Disk option in 1984, and later Tapeless Studio system.

Yamaha's GS-1, the first FM digital synthesizer, was released in 1980. To program the synthesizer, Yamaha built a custom computer workstation designed to be used as a sequencer for the GS-1. It was only available at Yamaha's headquarters in Japan (Hamamatsu) and the United States (Buena Park).

In 1982, renewed the Fairlight CMI Series II with its sequencer, "Page R", which combined step sequencing with sample playback.

Roland MC-8 Microcomposer (1977)

Standalone CV/Gate Sequencers

In 1977, Roland Corporation released the MC-8 Microcomposer, also called *computer music composer* by Roland. It was an early stand-alone, microprocessor-based, digital CV/Gate sequencer, and an early polyphonic sequencer. It equipped a keypad to enter notes as numeric codes, 16 KB of RAM for a maximum of 5200 notes (large for the time), and a polyphony function which allocated multiple pitch CVs to a single Gate. It was capable of eight-channel polyphony, allowing the creation of polyrhythmic sequences. While there were earlier microprocessor-based sequencers, they were based on keyboard entry, and lacked the MC-8's CV/Gate capabilities and depth of control/synchronization facilities. The MC-8 had a significant impact on popular electronic music, with the MC-8 and its descendants (such as the Roland MC-4 Microcomposer) impacting popular electronic music production in the 1970s and 1980s more than any other family of sequencers. The MC-8's earliest known users were Yellow Magic Orchestra in 1978.

MIDI Sequencers

In June 1981, Roland Corporation founder Ikutaro Kakehashi proposed the concept of standardization between different manufacturers' instruments as well as computers, to Oberheim Electronics founder Tom Oberheim and Sequential Circuits president Dave Smith. In October 1981, Kakehashi, Oberheim and Smith discussed the concept with representatives from Yamaha, Korg and Kawai. In 1983, the MIDI standard was unveiled by Kakehashi and Smith. The first MIDI sequencer was the Roland MSQ-700, released in 1983.

It was not until the advent of MIDI that general-purpose computers started to play a role as sequencers. Following the widespread adoption of MIDI, computer-based MIDI sequencers were developed. MIDI-to-CV/Gate converters were then used to enable analogue synthesizers to be controlled by a MIDI sequencer. Since its introduction, MIDI has remained the musical instrument industry standard interface through to the present day.

Personal Computers

In 1978, Japanese personal computers such as the Sharp MZ and Hitachi Basic Master were capable of digital synthesis, which were sequenced using Music Macro Language (MML). This was used to produce chiptune video game music.

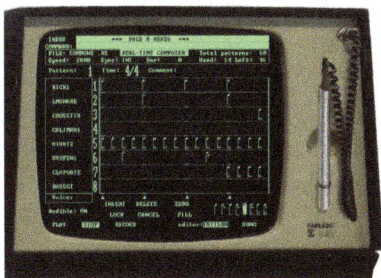

Page R on Fairligh

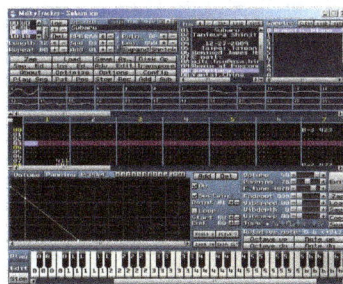

Tracker software

It was not until the advent of MIDI, introduced to the public in 1983, that general-purpose computers really started to play a role as software sequencers. NEC's personal computers, the PC-88 and PC-98, added support for MIDI sequencing with MML programming in 1982. In 1983, Yamaha modules for the MSX featured music production capabilities, real-time FM synthesis with sequencing, MIDI sequencing, and a graphical user interface for the software sequencer. Also in

1983, Roland Corporation's CMU-800 sound module introduced music synthesis and sequencing to the PC, Apple II, and Commodore 64.

The spread of MIDI on personal computers was facilitated by Roland's MPU-401, released in 1984. It was the first MIDI-equipped PC sound card, capable of MIDI sound processing and sequencing. After Roland sold MPU sound chips to other sound card manufacturers, it established a universal standard MIDI-to-PC interface. Following the widespread adoption of MIDI, computer-based MIDI software sequencers were developed.

In 1987, software sequencers called trackers were developed. They became popular in the 1980s and 1990s as simple sequencers for creating computer game music, and remain popular in the demoscene and chiptune music.

References

- J. Eargle and M. Gander (2004). "Historical Perspectives and Technology Overview of Loudspeakers for Sound Reinforcemen" (PDF). Journal of the Audio Engineering Society. 52 (4): 412–432 (p. 416)

- Rumsey, Francis; McCormick, Tim (2009). Sound and recording (6th ed.). Oxford, UK: Focal Press. p. 81. ISBN 978-0-240-52163-3

- Young, Tom (December 1, 2008). "In-Depth: The Aux-Fed Subwoofer Technique Explained". Study Hall. Pro-SoundWeb. p. 2. Archived from the original on January 14, 2010. Retrieved March 3, 2010

- Xiao, Lin; Kaili Jiang (2008). "Flexible, Stretchable, Transparent Carbon Nanotube Thin Film Loudspeakers". Nano Letters. 8 (12): 4539–4545. PMID 19367976. doi:10.1021/nl802750z

- Montgomery, Henry C (1959). "Amplification and High Fidelity in the Greek Theater". The Classical Journal. 54 (6): 242–245. JSTOR 3294133

- Eargle, John; Chris Foreman (2002). Audio Engineering for Sound Reinforcement. Milwaukee: Hal Leonard Corporation. p. 66. ISBN 0-634-04355-2

- DellaSala, Gene (August 29, 2004). "Setting the Subwoofer / LFE Crossover for Best Performance". Tips & Tricks: Get Good Bass. Audioholics. Retrieved March 3, 2010

- Wei, Yang; Xiaoyang Lin (2013). "Thermoacoustic Chips with Carbon Nanotube Thin Yarn Arrays". Nano Letters. 13 (10): 4795–801. PMID 24041369. doi:10.1021/nl402408j

- Sessler, G.M.; West, J.E. (1962). "Self-biased condenser microphone with high capacitance". Journal of the Acoustical Society of America. 34 (11): 1787–1788. doi:10.1121/1.1909130

- Rothstein, Joseph (1995). MIDI: A Comprehensive Introduction. Computer Music and Digital Audio Series. 7. A-R Editions, Inc. pp. 77, 122. ISBN 978-0-89579-309-6

- Zimmer, Ben (29 July 2010). "How Should 'Microphone' be Abbreviated?". The New York Times. Retrieved 10 September 2010

- Paritsky, Alexander; Kots, A. (1997). "Fiber optic microphone as a realization of fiber optic positioning sensors". Proc. of International Society for Optical Engineering (SPIE). 3110: 408–409. doi:10.1117/12.281371

- "Local firms strum the chords of real music innovation". Mass High Tech: the Journal of New England Technology. February 8, 2008

- Pinch, Trevor. J.; Trocco, Frank (2009). "Buchla's Box". Analog Days: The Invention and Impact of the Moog Synthesizer (reprint ed.). Harvard University Press. pp. 55–56. ISBN 978-0-674-04216-2

- Dave Berners (December 2005). "Ask the Doctors: The Physics of Mid-Side (MS) Miking". Universal Audio WebZine. Universal Audio. Retrieved 2013-07-30

- Fowler, Charles B. (October 1967). "The Museum of Music: A History of Mechanical Instruments". Music Educators Journal. Music Educators Journal. 54 (2): 45–49. JSTOR 3391092. doi:10.2307/3391092

- Bogdanov, Vladimir (2001). All music guide to electronica: the definitive guide to electronic music. Backbeat Books. pp. 320. ISBN 978-0-87930-628-1. Koetsier, Teun (2001). "On the prehistory of programmable machines: musical automata, looms, calculators". Mechanism and Machine Theory. Elsevier. 36 (5): 589–603. doi:10.1016/S0094-114X(01)00005-2

4

Audio Mixing: Tools and Techniques

The act of mixing several tracks or recordings is known as audio mixing. The techniques used in the process mainly depend upon the quality and particular genres the recordings belong to. The chapter serves as a source to understand the major categories related to audio mixing.

Audio Mixing (Recorded Music)

Digital Mixing Console Sony DMX R-100 used in project studios

In sound recording and reproduction, audio mixing is the process of combining multitrack recordings into a single track and these tracks that are blended together are done so by using various processes such as EQ, Compression and Reverb. The track may be mixed in mono, stereo, or surround sound. There are numerous approaches, methods and techniques involved in Audio mixing; some of these practices include levels setting, equalization, stereo panning, and effects. Audio mixing techniques and approaches can vary widely, and these can greatly affect the qualities of the sound recording.

Audio mixing techniques largely depend on music genres and the quality of sound recordings involved. The process is generally carried out by a mixing engineer, though sometimes the musical producer or music artist may assist. After mixing, a mastering engineer prepares the final product for production.

Audio mixing may be transferred onto a mixing console or digital audio workstation.

History

Early Recording Machines

In the late nineteenth century, Thomas Edison and Emile Berliner developed the first recording machines. The recording and reproduction process itself was completely mechanical with little or no electrical parts. Edison's phonograph cylinder system utilized a small horn terminated in a stretched, flexible diaphragm attached to a stylus which cut a groove of varying depth into the malleable tin foil of the cylinder. Emile Berliner's gramophone system recorded music by inscribing spiraling lateral cuts onto a vinyl disc.

Electronic recording became more widely used during the 1920s. It was based on the principles of electromagnetic transduction. The possibility for a microphone to be connected remotely to a recording machine meant that microphones could be positioned in more suitable places. Even more useful was the fact that the outputs of the microphones could be mixed before being fed to the disc cutter, allowing greater flexibility in the balance.

Before the introduction of multitrack recording, all sounds and effects that were to be part of a record were mixed at one time during a live performance. If the recorded blend (or mix, as it is called) wasn't satisfactory, or if one musician made a mistake, the selection had to be performed over until the desired balance and performance was obtained. However, with the introduction of multi-track recording, the production phase of a modern recording has radically changed into one that generally involves three stages: recording, overdubbing, and downmix.

Modern mixing emerged with the introduction of commercial multi-track tape machines, most notably the 8-track recorders that were introduced during the 1960s. The ability to record sounds into a multitude of channels meant that treating these sounds could be postponed to a later stage—the mixing stage.

In the 1980s, home recording and mixing became much easier. The 4-track Portastudio was introduced in 1979. Bruce Springsteen released the album *Nebraska* in 1982 using one. The Eurythmics topped the charts in 1983 with the song "Sweet Dreams (Are Made of This)", recorded by band member Dave Stewart on a makeshift 8-track recorder. In the mid-to-late 1990s, computers replaced tape-based recording for most home studios, with the Power Macintosh proving popular. At the same time, digital audio workstations, first used in the mid-1980s, began to replace tape in many professional recording studios.

Equipment

Mixing Consoles

A mixer (mixing console, mixing desk, mixing board, or software mixer) is the operational heart of the mixing process. Mixers offer a multitude of inputs, each fed by a track from a multitrack recorder. Mixers typically have 2 main outputs (in the case of two-channel stereo mixing) or 8 (in the case of surround).

Mixers offer three main functionalities:

- Mixing – summing signals together, which is normally done by a dedicated summing amplifier or in the case of digital by a simple algorithm.

- Routing – allows the routing of source signals to internal buses or external processing units and effects.

- Processing – many mixers also offer on-board processors, like equalizers and compressors.

A simple mixing console

Mixing consoles used for dubbing can often be seen as large and intimidating, due to the exceptional amount of controls. However, because many of these controls are duplicated, much of the console can be learnt by studying one part of it. The controls on a mixing console will typically fall into one of two categories: processing and configuration. Processors are the controls used to manipulate the sound. These can vary in complexity, from simple internal level controls, to sophisticated outboard reverberation units. Configuration controls deal with the signal routing from the input to the output of the console through the various processes.

Digital audio workstations (DAW) have many mixing features which potentially have more processes available than that of a major console. The distinction between a large console and a DAW equipped with a control surface is that a digital console will typically consist of dedicated digital signal processors for each channel. It is thus designed not to "overload" under the burden of signal processing, which may crash or lose signals. DAWs can dynamically assign resources like digital audio signal processing power, but may run out if too many signal processes are in simultaneous use. This overload can be solved fairly easily by simply plugging more hardware into the DAW, although the cost of such an endeavour may begin to approach that of a major console.

Outboard Gear and Plugins

Outboard gear (analogue) and software plugins (digital) can be inserted into the signal path to extend processing possibilities. Outboard gear and plugins fall into two main categories:

- Processors – these devices are normally connected in series to the signal path, so the input signal is replaced with the processed signal (e.g. equalizers).

- Effects – these can be considered as any unit that has an effect upon the signal, the term is mostly used to describe units that are connected in parallel to the signal path, and therefore they add to the existing sounds but do not replace them. Examples would include reverb and delay.

Multiple Level Controls in Signal Path

A single signal can pass through a large number of level controls – such as an individual channel fader, subgroup master fader, master fader and monitor volume control. According to audio engineer Tomlinson Holman, problems are created due to the multiplicity of the controls. Each and every console has their own dynamic range and it is important to utilize this correctly to avoid excessive noise or distortions. Attacking this problem – of the correct setting for the variety of controls - can be accomplished relatively quickly. Holman refers to the scale of the control as a clue for the solution of this problem. With 0 dB being the nominal setting of the controls, many have a "gain in hand," which goes above 0 dB. This means that one can turn it up from the nominal setting to have something that sounds clear. Other controls, such as sub masters and master level controls, are used for slight trims to the overall section-by-section balance or for the main fade-ins and fade-outs of the overall mix.

Processes that Affect Levels

- Faders – used to attenuate or boost the level of signals.

- Pan pots – A fundamental part of configuration in recording console is panning. Pan pots are devices that place sound among the channels: L, C, R, LS, and RS. They are also used to pan signals to the left or right and in surround, to the back or front.

- Compressors – A device which automatically varies the volume range of tracks being mixed, so that one track is not obscured by another when a low volume level on the primary track coincides with a high volume level on a secondary track. Compressors are equipped with a number of controls to vary the volume range over which the action of compression occurs, the amount of compression, and how quickly or slowly the compressor acts.

- Expansion – The Expansion device does exactly the opposite of what the compressor does. It increases the volume range of a source and may do so across a wide dynamic range or may be restricted to a narrower region by control functions. Restricting expansion to only low-level sounds helps to minimize noise. This function is often referred to as downward expansion, noise gating, or keying and reduces the level below a threshold set by a specific control. Noise gates have numerous audible problems. (e.g.: In a dialog recording with air conditioning noise in the background, the threshold of the noise gate may remove the air conditioner sound between lines of dialog which can create an exaggerated difference that could be much more noticeable than if the audio had been left unprocessed.)

- Limiters – A limiter acts on signals above a certain threshold. Above that threshold, the level is controlled so that for each dB of increase on the input, the gain is reduced by the same amount. Therefore, the output level above the threshold will stay exactly the same, regardless of any increases in the input level. Limiters can be used to catch occasional events that might not otherwise be controlled, to bring them into a range in which the recording medium can handle the signal linearly.

These items discussed thus far affect the level of audio signal. The most commonly used process is level control, which is used even on the simplest of mixers.

Processes that Affect Frequency Response

Processes that primarily affect the frequency response of the signal are generally seen as second in importance to level control. These processes clean the audio signal, enhance interchangeability between other signals, adjust for the loudness effect, and generally create a much more pleasant or deliberately worse sound. There are two principle frequency response processes – equalization and filtering.

- Equalizers – The simplest description of EQ is the process of altering the frequency response in a manner similar to what tone controls do on a stereo system. Professional EQs dissect the audio spectrum into three or four parts which may be called the low-bass, mid-bass, mid-treble, and high frequency controls.

- Filters – Filters are used to essentially eliminate certain frequencies from the output. Filters strip off the any part of the audio spectrum. There are various types of filters. A high-pass filter (low-cut) is used to remove excessive room noise at low frequencies. A low-pass filter (high-cut) is used to help isolate a low frequency instrument playing in a studio along with others. And a band-pass filter is a combination of high- and low-pass filters, also known as a telephone filter (because a sound lacking in high and low frequencies resembles the quality of sound transmitted and received by telephone).

Processes that Affect Time

- Reverbs – Reverbs are used to simulate boundary reflections created in a real room, adding a sense of space and depth to otherwise ‹dry› recordings. Another use is to distinguish among auditory objects; all sound having one reverberant character will be categorized together by human hearing in a process called auditory streaming. This is an important feature in layering sound, in depth, from in front of the speaker to behind it.

Before the advent of electronic reverb and echo processing, physical means were used to generate the effects. An echo chamber, a large reverberant room, could be equipped with a speaker and at least two spaced microphones. Signals were then sent to the speaker and the reverberation generated in the room was picked up by the two microphones, constituting a "stereo return".

Downmixing

Downmixing is the process of converting a program with a multiple-channel configuration into a program with fewer channels. Common examples include downmixing from 5.1 surround sound to stereo, and stereo to mono. In the former case, the left and right surround channels are blended with the left and right front channels. The centre channel is blended equally with the left and right channels. The LFE channel is either mixed with the front signals or not used. Because these are common scenarios, it is common practice to verify the sound of such downmixes during the production process to ensure stereo and mono compatibility.

The alternative channel configuration can be explicitly authored during the production process with multiple channel configurations provided for distribution. For example, a stereo mix can be put on DVDAudio discs or Super Audio CDs along with the surround mix. Alternatively, the program can be automatically downmixed by the end consumer's audio system. For example, a DVD player or sound card may downmix a surround sound program to stereophonic sound (two channels) for playback through two speakers.

Mixing in Surround Sound

Any device having a number of multiple bus consoles (typically having eight or more buses) can be used to create a 5.1 surround sound mix, but this may be frustrating if the device is not designed to facilitate signal routing, panning and processing in a surround sound environment. Whether working in an analog hardware, digital hardware, or DAW "in-the-box" mixing environment, the ability to pan mono or stereo sources and place effects in the 5.1 soundscape and monitor multiple output formats without difficulty can make the difference between a successful or compromised mix. Mixing in surround is very similar to mixing in stereo except that there are more speakers, placed to "surround" the listener. In addition to the horizontal panoramic options available in stereo, mixing in surround lets the mix engineer pan sources within a much wider and more enveloping environment. In a surround mix, sounds can appear to originate from many more or almost any direction depending on the number of speakers used, their placement and how audio is processed.

There are two common ways to approach mixing in surround:

- Expanded Stereo – With this approach, the mix will still sound very much like an ordinary stereo mix. Most of the sources such as the instruments of a band, the vocals, and so on, will still be panned between the left and right speakers, but lower levels might also be sent to the rear speakers in order to create a wider stereo image, while lead sources such as the main vocal might be sent to the center speaker. Additionally, reverb and delay effects will often be sent to the rear speakers to create a more realistic sense of being in a real acoustic space. In the case of mixing a live recording that was performed in front of an audience, signals recorded by microphones aimed at, or placed among the audience will also often be sent to the rear speakers to make the listener feel as if he or she is actually a part of the audience.

- Complete Surround/All speakers are treated equally – Instead of following the traditional ways of mixing in stereo, this much more liberal approach lets the mix engineer do anything he or she wants. Instruments can appear to originate from anywhere, or even spin around the listener. When done appropriately and with taste, interesting sonic experiences can be achieved, as was the case with James Guthrie's 5.1 mix of Pink Floyd's *The Dark Side of the Moon*, albeit with input from the band. This is a much different mix from the 1970s quadrophonic mix.

Naturally, these two approaches can be combined any way the mix engineer sees fit. Recently, a third approach to mixing in surround was developed by surround mix engineer Unne Liljeblad.

- MSS – Multi Stereo Surround – This approach treats the speakers in a surround sound system as a multitude of stereo pairs. For example, a stereo recording of a piano, created using two microphones in an ORTF configuration, might have its left channel sent to the left rear speaker and its right channel sent to the center speaker. The piano might also be sent to a reverb having its left and right outputs sent to the left front speaker and right rear speaker, respectively. Additional elements of the song, such as an acoustic guitar recorded in stereo, might have its left and right channels sent to a different stereo pair such as the left front speaker and the right rear speaker with its reverb returning to yet another stereo pair, the left rear speaker and the center speaker. Thus, multiple clean stereo recordings surround the listener without the smearing comb-filtering effects that often occur when the same or similar sources are sent to multiple speakers.

Automixer

Microphones at a press conference being processed through a Dugan E-1 automixer which has been placed on top of the regular audio mixer. San Francisco mayor Gavin Newsom is speaking at a lectern, while golfers Fred Couples and Greg Norman are seated on stage. Six of eight automixer inputs have been muted and are showing red LEDs. The active input is showing full gain with a ladder of green LEDs

An automixer, or automatic microphone mixer, is a live sound mixing device that automatically reduces the strength of a microphone's audio signal when it is not being used. Automixers lower the hiss, rumble, reverberation and other extraneous noise that occur when several microphones operate simultaneously. They may also be used to mix sound from non-microphone signals such as playback devices.

Automixers are typically used to mix panel discussions on television talk shows and at conferences and seminars. They can also be used to mix actors' wireless microphones in theater productions and musicals. Automixers are frequently employed in settings where it is expected that a live sound operator will be not present, such as courtrooms and city council chambers.

Automixer hardware or software uses a variety of methods that allow increased gain before feedback for live sound reinforcement as well as reducing comb filtering between multiple microphones for recorded and broadcast applications.

Use

Automixers balance multiple sound sources based on each source's level, quickly and dramatically adjusting the various signal levels automatically. Automixers are used in live sound reinforcement to maintain a steady limit on the overall signal level of the microphones; if a public address system is set up so that one microphone will not feed back, then, in general, multiple microphones will not feed back if they are automixed. The equivalent number of open mics (NOM) present at the output of the automixer is kept low, regardless of the actual number of open mics.

While a skilled audio mix operator can greatly enhance the performance of a sound reinforcement system he cannot anticipate with perfect accuracy which participant will speak next in a spontaneous discussion. Sudden interjections by panelists may be lost completely, or the beginning of a word may be absent (or *upcut*) because the operator does not respond quickly enough to "fade up" an audio signal. A properly adjusted automixer can prevent losing words or phrases due to upcut mistakes or lapses of attention.

Priority Ducking is used to lower the level of a microphone(s) and/or program material based on a source signal from another microphone, for the duration of the signal present at the source microphone. It restores the original level once the source signal has ceased. This is useful when a) program material needs to attenuate in order to accentuate the voice of a narrator, b) one microphone is used by a chairman or master of ceremonies, and needs to have priority over other mics and/or program material, or c) a paging mic must attenuate all other signals.

Automixers can be oriented toward live reinforcement application or permanent installation with varying degrees of connector type (XLR or block) and feature switches and channel potentiometers located internally or externally. Channel switches permit the user to vary sensitivity (dynamic mic, condenser mic, or line), phantom power, priority, and always on-or-auto mode ("on" being useful for uninterrupted music playback). Auto mixers are frequently configured in board rooms, but are also useful in theatrical and "town hall" events where several wireless microphones are in use. Auto mixers may be connected directly to wired mics and wireless mic receivers, or may be connected to the channel direct output or send jacks of a mixing console. Integration of an auto mixer with manual mixing console permits more precise input gain control, absolute mic muting, low cut filter, equalization and individual mic channel foldback control.

History

Frank J. Clement and Bell Labs received a patent in 1969 for a multiple station conference telephone system that switched its output to the loudest input. The next year, Emil Torick and Richard G. Allen were granted a patent for an "Automatic Gain Control System with Noise Variable Threshold", an adaptive threshold circuit invention with its patent assignation going to Columbia Broadcasting System.

Some systems using electro-mechanical switching in regard to microphone activation were engineered in the late 1960s and early 1970s. Peter W. Tappan and Robert F. Ancha devised a system of seat sensors that would activate one of 350 hidden microphones at the Seventeenth Church of Christ, Scientist in Chicago in 1970. From approximately 1968, Ken Patterson and Diversified Concepts developed a hardware system that could detect the "Number of Open Microphones" (NOM) and attenuate the master output by an amount which increased with a higher number of microphones in use. This latter system was public domain.

In 1971, Gregory Maston of Bell Labs filed for a patent involving a circuit that could switch between several audio sources based on their levels. The loudest one was latched into the mix. This system did not ramp switched signals smoothly in and out and did not maintain a constant ambience. It was intended for speakerphone conferencing applications. In 1972, Keith A. T. Knox with the British Post Office Corporation developed an adaptive threshold gate circuit intended for speakerphone usage. The system used a second microphone somewhat near the first to sense ambient noise level.

Dan Dugan showed his first "Adaptive Threshold Automatic Microphone Mixing System" in 1974 at the 49th Audio Engineering Society (AES) meeting in New York, and was granted a patent for a control apparatus for sound reinforcement systems which sensed ambient sound level in the environment of a theater to control each microphone's individual level. In 1976, Dugan was granted a patent for an automatic microphone mixing process whereby the total gain of the system remains constant. He began manufacturing his first automixer system, the Model A, based on his two patents. Dugan built 60 units, with the first, hand-assembled one taken to Bell Labs to be installed in their conference room for Harvey Fletcher. The algorithm was simple and effective: *Each individual input channel is attenuated by an amount, in dB, equal to the difference, in dB, between that channel's level and the sum of all channel levels.* Dugan licensed the system to Altec who released several automixer models including the 1674A, -B and -C series and the 1684A, all focusing on speech applications. (The 1684A became an Electrovoice product and is currently administered by their Commercial division.) The earliest Altec product implementation was regarded as inferior within the commercial audio contractor industry, and other manufacturers began to design their own automixer products.

Dan Dugan's first automixers

In 1978, Richard W. Peters of Industrial Research Products (IRP) was granted an improvement patent entitled "Priority mixer control". IRP released the Voice-Matic series of 4 × 1 and 8 × 1 automatic mixers using "Dynamic Threshold Sensing" that weighed a combination of the amplitude and history of the signal to determine channel access. The master output was attenuated at the rate of 3 dB for every doubling of NOM. This master output reduction was the solution used by Yamaha Pro Audio two decades later in their DME series of digital signal processing (DSP) products, incorporating an automixer function which was otherwise an 8- or 16-channel noise gate.

Eugene Campbell and Terrance Whittemore of Colorado were granted a patent in 1982 for an automatic microphone mixing algorithm that allowed for musical performance mixing that would not be dominated by the loudest vocalist or instrumentalist.

Graphic user interface for a digital automixer that uses a gain-sharing algorithm. Controls include threshold, depth, polarity inversion and muting for each input, as well as volume controls for the four inputs, the four individual outputs and the full mix output

Stephen D. Julstrom of Shure Brothers, Inc. (Evanston, Illinois) was granted a patent in 1987 for a teleconferencing system that used special directionally gated microphones mixed automatically and sent to a distant party via telephone line. The return signal from the distant party was compared in strength to the local mix to determine which party was to be most prominent in the overall mix. Any interrupting party was given priority. Four years later, Shure would introduce the AMS4000 and AMS8000 automixers for sound reinforcement; mixers which required the use of special directional condenser microphones of the Shure AMS Series.

In 1985 Innovative Electronic Designs (IED) introduced the circuit card frame-based auto mic mixing system featuring combine-separate function and programmable gain control (PGC) modules. Combine-separate functionality is useful for ballroom applications with movable partitions that permit portions or all of a room to be used for one program or multiple programs. PGC modules compensate individual mic channel gain for soft-spoken or loud presenters, by slightly boosting or cutting channel gain, respectively.

At the 87th AES Convention in 1989, Dugan introduced the idea of using an automixer inserted within target channels on a professional audio mixer. Each microphone's signal would be interrupted inside its channel strip and sent to a variable gain circuit contained within the automixer. The signal would then be returned to the mixer at a level consistent with the Dugan algorithm. This became the Dugan Model D automixer.

In 1991, Dugan's patent expired. Competing manufacturers began to bring the Dugan algorithm directly to their product designs. In 1993, Travis M. Sims, Jr. of Lectrosonics (Rio Rancho, New Mexico) was granted a patent for a sound system with rate controlled, variable attenuation of microphone inputs, including the Dugan algorithm as well as loudspeaker zone attenuation when in close proximity to an active microphone. The loudspeaker zone part of the patent cited a 1985 patent for proportional amplification by Eugene R. Griffith, Jr. of LVW Systems of Colorado Springs, a commercial audio contractor. In 1995, Sims and Lectrosonics gained another patent for an "Adaptive proportional gain audio mixing system" which incorporated a number of ideas including the Dugan algorithm for maintaining a constant total gain of all the inputs.

In 1996, Dugan came out with the Model D-1, a speech-only economy model that did not offer the music system of the Model D.

In 1997, John H. Roberts of Peavey Electronics was granted a patent for an automatic mixer priority circuit, enabling a hierarchy of logic weighting that allowed selected signals to push forward in the mix when they are in use, while still maintaining the useful constant unity, gain-sharing relationship first described by Dugan. The hierarchy enabled a host, moderator or chairperson to speak over other participants and retain control of the discussion. Peavey's Architectural Acoustics division used three levels of hierarchy in their 1998 "Automix 2" product, placing the first- and second-most influentially weighted sources at inputs 1 and 2, respectively.

Dan Dugan licensed his system to Protech Audio (Indian Lake, New York) in 1997, yielding the Protech 2000 model series.

In 2004, the first standard audio mixer incorporating an eight-channel automixer section was released by Peavey in their Sanctuary Series, and in 2006 the similar HP-W was introduced by Crest.

Both mixers were aimed at the House of Worship market, adding functions that ease the process of audio mixing to religious organizations.

In 2007, Mark W. Gilbert and Gregory H. Canfield of Shure (Niles, Illinois) were granted a patent for a digital microphone automixer system that used time of arrival as its main decision-making criteria.

Generations of Dugan's insertable automixers

In February 2011, Dugan announced an automixer card to plug in the accessory slot of a Yamaha digital mixing console such as the LS-9, M7CL or PM-5D. This card, the Dugan-MY16, can mix 16 channels of microphone inputs at 44.1–48 kHz or 8 channels at 88.2–96 kHz sampling rates. Channels to be automixed are assigned in the mixer's graphic user interface, and can then be controlled by a common web browser interface affecting only the Dugan-MY16 card, allowing remote control with an iPad, touchscreen computer or laptop over wireless network.

Related Applications

- Speech intelligibility enhancement, James M. Kates of Signatron (1984). This system uses Dugan's automatic mixing algorithm to reconstitute several spectral regions of a signal that has been divided into frequency bands for short-time spectral analysis in order to achieve greater intelligibility of spoken consonants.

- Secure conferencing, patent by Raoul E. Drapeau (1993). An automixing algorithm attempts to mask incidental speech that is below automix threshold but which can be audible in the mix. The automix circuitry indicates which sources are active, and whether masking of low-level signals is occurring.

Matrix Mixer

A matrix mixer is an audio electronics device that routes multiple input audio signals to multiple outputs. It usually employs level controls such as potentiometers to determine how much of each input is going to each output, and it can incorporate simple on/off assignment buttons. The number of individual controls is at least the number of inputs multiplied by the number of outputs.

A 19-inch rack holding several professional audio devices including an 8×8 matrix mixer at the bottom, made by Midas Consoles. The matrix mixer has 8 vertical faders to control output level, 8 light gray potentiometers (rotating pots) for input level control, and 64 dark gray pots for matrix mixing. There are also 64 on/off buttons, one for each input/output intersection.

Matrix mixers may be incorporated into larger devices such as mixing consoles or they may be a standalone product. They always have routing and level controls and may also include other features. Matrix mixers are often used in a complex listening space to send audio signals to different loudspeaker zones. They may be used to provide the producer or director different blends of a mixing project for television, film or recording studio.

Basics

In professional audio, a matrix mixer is used to route audio signals from multiple sources to different destinations. It may be a standalone device or embedded within another, larger product such as a mixing console, digital audio workstation or digital signal processor.

An analog matrix mixer contains a number of simple electronic mixer circuits, the quantity being the number of inputs multiplied by the number of outputs. Each electronic mixer controls the level (gain) of one input going to one output. The level control is usually a rotating potentiometer (called a "pot"). Each row of electronic mixer circuits, one from each input, feeds a summing amplifier or combining amp at the output. A fader (a linear potentiometer) may be used to control the level of each output signal. Other controls may include a mute button for each input/output intersection, a mute button for each input, a mute button for each output, and buttons that invert the input signal polarity. The output signals of the matrix mixer may be digital, or they may be balanced or unbalanced analog. It is possible in a matrix mixer to combine an all-analog signal path with digital control of level.

Any audio console that lacks a matrix section may be connected to an external matrix mixer. Many audio manufacturers have produced matrixes incorporating digital signal processing (DSP) which offers additional tools such as compression, equalization, ducking, gating, and loudspeaker system management. A DSP-based matrix system may be optimized by expert audio engineers then made inaccessible to inexperienced soundboard operators.

Live Sound

In live sound mixing the front of house mixing console often includes a matrix mixer section for optimizing the blend of sound going to various locations in the performance space. Subgroups, auxiliary mixes, submixes or even the main outputs are routed through the matrix mixer to different loudspeaker zones, making the matrix essentially a "mix of mixes". A front-of-house matrix may be used at a concert to switch between the headliner's and the opening act's mixing consoles. A matrix mixer feature may be included on analog or digital consoles. For instance, the Midas Consoles XL4 concert mixing console has a built-in 8×8 matrix summing 8 subgroups into 8 matrix outputs; the same basic module is available separately as the Midas XL88. The digital Yamaha M7CL mixing console is another such product; it has a 19×8 matrix section summing 16 mixes as well as the main left/right/mono outputs. In the classic analog monitor mixer without faders, the console is set up as a large matrix mixer with the addition of equalization filters and other controls for each input.

Yamaha M7CL 19×8 matrix section, showing the 16 mix buses plus the main L/R/Mono buses routed to 8 matrix outputs. The individual level controls are shown as horizontal bars in dark orange. The 8 matrix outputs are controlled by 8 white faders at the bottom.

The matrix mixer is useful for performance spaces with multiple loudspeaker zones such as main left and right loudspeaker arrays, a center loudspeaker cluster or array, under-balcony speakers for the rear seating area, stage-lip loudspeakers for the front seats, overflow rooms, green rooms, the foyer, and broadcast or live streaming to distant audiences. If there is no separate monitor mixer with its own operator, the matrix mixer could be used to send a different blend to stage monitors or sidefill loudspeakers. However, such a use of the matrix is limited by the likelihood that the musicians on stage would be disturbed by any changes made to the channel levels by the front of house operator. It is usually better to supply stage monitors with signal obtained from pre-fader auxiliary sends. Listeners outside of the main performance space will not hear amplified instruments such as electric guitar as well as those inside, so the matrix output going to their location would typically contain more of the amplified instruments. Also, the matrix may be used to send a different blend to any recording devices, especially a blend that is optimized for stereo playback whereas the live sound for the performance space may be presented in mono. The matrix mixer may be used to combine the main mix with audience ambiance microphones to obtain a more "live" blend for the recording mix.

Recorded Sound

In the process of mixing for film or television, a matrix mixer may be used to give the film director, television director or producer a working mix of the project while the mix engineer puts it together.

In the recording studio, the same method may be used to give the record producer a different blend during the mixing process. As well, the matrix mixer may have been used to route various audio input channels to specific recording channels

Digital Mixing Console

In professional audio, a digital mixing console (DMC) is an electronic device used to combine, route, and change the dynamics, equalization and other properties of multiple audio input signals, using digital computers rather than analog circuitry. The digital audio samples, which is the internal representation of the analog inputs, are summed to what is known as a master channel to produce a combined output. A professional digital mixing console is a dedicated desk or control surface produced exclusively for the task, and is typically more robust in terms of user control, processing power and quality of audio effects. However, a computer with proper controller hardware can act as the device for the digital mixing console since it can mimic its interface, input and output.

Yamaha M7CL in place for a live production.

Uses

Digital mixing consoles are typically used in recording studios, public address systems, sound reinforcement systems, broadcasting, television, and film post-production.

Common Sound System Problems and Solutions

Most DMCs are expensive and sophisticated tools. The most common issue related to the DMC is the complex structure, which can be difficult to navigate without previous experience or knowledge of the system. The user usually requires a basic understanding of signal flow, audio terminology, and hardware implementation.

Part of the solution to alleviate operator issues is to automate whenever possible. The advent of modern digital computer technology has now made it possible to install sound system components that will almost operate themselves. When upgrading or installing new sound systems, it is advantageous to install items that require as little hands-on, human operation as possible.

A digital mixing console can offset the lack of operator experience because it can store the settings programmed by an expert mixer. For example: A knowledgeable person can adjust all of the microphone settings, monitors, etc. for a given event. After everything is properly adjusted, that set-up is assigned a name and stored in the memory. Afterwards, a less knowledgeable operator can simply select that setting on their console or computer.

One can easily program many different preset configurations or "snapshots," into the mixing console. Default configurations that are included with the DMC are known as presets. Once a stored setting is recalled, the operator can still make manual volume adjustments, etc., without affecting the stored program. In other words, they can change a lot of stuff and all one has to do is hit the recall and the mixer automatically returns to all of the correct start-up settings.

Another pitfall in terms of live applications is improper location of the equipment. No sound operator can properly adjust a live sound system unless he can hear exactly what the majority of the audience hears, yet for issues related to space, appearance and security, one cannot always locate their sound control equipment in the middle of their auditorium. This is even more of a problem if the auditorium is a multi-use building that is often converted for other events. A digital mixer may solve this problem: a sound operator can operate the whole sound system from a laptop computer. With the proper set-up, it can even be done by a wireless tablet for increased mobility. In fact, many of the digital mixer's functions are easier to operate from a computer screen than the actual mixing console.

Digidesign's Venue Profile mixer on location at a corporate event. This mixer allows plugins from third-party vendors. A Smaart software screen is partially shown on the right—Smaart allows plugins, too.

Another advantage of DMCs is the abundance of control features that it provides for each input channel. The built-in effects of typical DMCs are robust and include gates, compressor limiters, equalizers, feedback controllers, and other signal processing hardware. One advantage of the large number of internal effects is that a DMC system is less vulnerable to failure and outside interference than a set-up using outboard hardware.

Many digital mixers have controls that mimic the classic look and feel of analog mixers. This is comparable to a fly-by-wire system in modern aircraft. The controls are similar, but the underlying mechanism has changed from voltage levels to binary information.

Third-party plug ins can add functionality in a digital mixer. Plugins allow for further expansion of the mixer's on-board equalization, compression and reverberation effects.

Using Dual DMCs to Improve Live Recording

In truly professional broadcast and recording applications, one does not use what is referred to as the house mix for high quality audio recordings. The reason for this is that when engineering live sound for any auditorium, one must deal with the acoustic parameters of that particular auditorium. This requires various adjustments of equalization, bass, treble, volume, etc. While those adjustments may enhance the sound quality in the auditorium, they are not necessarily needed for the recording. In fact, those house mix adjustments often diminish the quality of the recorded sound. Once the bass, treble, volume and other effects of the house mix are added to the recording mix, it is most difficult to correct. In the reverse, adjustments and signal processing effects that are often used to enhance a recording mix are not always needed in the house mix.

In order to facilitate this, the signals must be split and provided to OB, recording crew or film crews. This split can be analog or digital; an analog split will normally be a feed of all stage signals split either passively or through an isolated transformer split. Transformers are preferred as they provide isolation and prevent microphone preamps on different systems interacting, for example a recording pre-amp causing level change at FOH or Monitors. A digital split can come in many forms, Often AES/EBU, MADI, Firewire direct to DAW, or increasingly network aware digital snake formats, such as AVB, Dante, Ethersound or Rocknet, many other proprietary formats also exist.

Advantages and Disadvantages

Advantages

- There is no added noise, (unintentional) distortion, or other signal degradation while the signal is in the digital domain, between the output of the analog to digital converter (ADC) and the input to the digital to analog converter (DAC).

- Aux sends can be mixed on the main faders rather than on a row of potentiometers.

- Signal routing is often much more flexible than with an analog-based console.

- The setup of the console can be saved and loaded at will. This is particularly useful in live events where a setup for each band can be largely prepared in advance, saved, and then loaded as needed.

- There are typically many on-board effects and virtual signal processors available, eliminating the need for additional hardware modules, and the associated cost, size, weight, cabling, signal quality issues, etc.

Disadvantages

- There is an analog-to-digital conversion, then processing of the signal, then a digital to analog conversion, which degrades the sound quality. This is subject to debate, since the quality degradation is not always noticeable.

- The number of faders is often fewer than the number of input channels. The extra input channels are not accessible until a bank of faders is switched to control them.

- Digital conversion and processing adds latency, or delay, into the signal.

- The act of making adjustments is often slower for compact digital mixers which require the user to page through one or more layers of commands before reaching the desired control.

Electronic Mixer

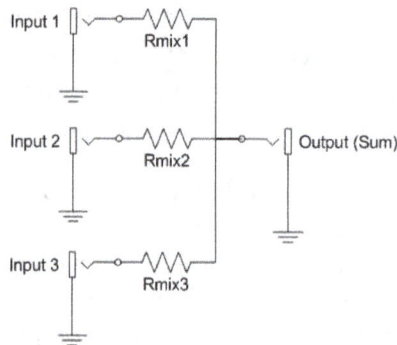

A simple three-channel passive additive mixer. More channels can be added
by simply adding more input jacks and mix resistors.

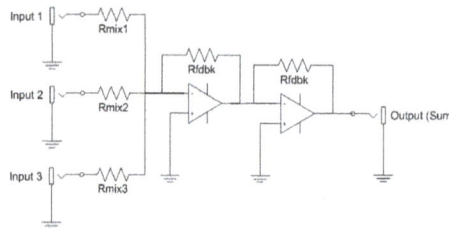

A "virtual ground" active additive mixer. The buffer amplifiers serve to reduce crosstalk and distortion.

An electronic mixer is a device that combines two or more electrical or electronic signals into one
or two composite output signals. There are two basic circuits that both use the term *mixer*, but they
are very different types of circuits: additive mixers and multiplicative mixers.

Simple additive mixers use Kirchhoff's circuit laws to add the currents of two or more signals together, and this terminology ("mixer") is only used in the realm of audio electronics where audio mixers
are used to add together audio signals such as voice signals, music signals, and sound effects.

Multiplicative mixers multiply together two time-varying input signals instantaneously (in-
stant-by-instant). If the two input signals are both sinusoids of specified frequencies f_1 and f_2, then
the output of the mixer will contain two new sinsoids that have the sum $f_1 + f_2$ frequency and the
difference frequency absolute value $|f_1 - f_2|$.

Note: Any nonlinear electronic block driven by two signals with frequencies f_1 and f_2 would generate
intermodulation (mixing) products. A multiplier (which is a nonlinear device) will generate ideally
only the sum and difference frequencies, whereas an arbitrary nonlinear block would generate also
signals at e.g. $2 \cdot f_1 - 3 \cdot f_2$, etc. Therefore, normal nonlinear amplifiers or just single diodes have been
used as mixers, instead of a more complex multiplier. A multiplier has usually the advantage of
rejecting - at least partly - undesired higher-order intermodulations and larger conversion gain.

Additive Mixers

Additive mixers add two or more signals, giving out a composite signal that contains the frequency components of each of the source signals. The simplest additive mixers are simple resistor

networks, and thus purely passive, while more complex matrix mixers employ active components such as buffer amplifiers for impedance matching and better isolation.

Multiplicative Mixers

Ideal multiplicative mixers produce an output signal equal to the product of the two input signals. Multiplicative mixers are often used in conjunction with an oscillator in the communications field to modulate signal frequencies. Multiplicative mixers can be coupled with a filter to either up-convert or down-convert an input signal frequency, but they are more commonly used to down-convert to a lower frequency to allow for simpler filter designs, as done in superheterodyne receivers. In many typical circuits, the single output signal actually contains multiple waveforms, namely those at the sum and difference of the two input frequencies and harmonic waveforms. The output signal may be obtained by removing the other signal components with a filter.

Mathematical Treatment

The received signal can be represented as

$$E_{sig} \cos(\omega_{sig}t + \varphi)$$

and that of the local oscillator can be represented as

$$E_{LO} \cos(\omega_{LO}t).$$

For simplicity, assume that the output I of the detector is proportional to the square of the amplitude:

$$I \propto \left(E_{sig} \cos(\omega_{sig}t + \varphi) + E_{LO} \cos(\omega_{LO}t) \right)^2$$

$$= \frac{E_{sig}^2}{2} \left(1 + \cos(2\omega_{sig}t + 2\varphi) \right)$$

$$+ \frac{E_{LO}^2}{2} \left(1 + \cos(2\omega_{LO}t) \right)$$

$$+ E_{sig}E_{LO} \left[\cos((\omega_{sig} + \omega_{LO})t + \varphi) + \cos((\omega_{sig} - \omega_{LO})t + \varphi) \right]$$

$$= \underbrace{\frac{E_{sig}^2 + E_{LO}^2}{2}}_{constant\ component} + \underbrace{\frac{E_{sig}^2}{2}\cos(2\omega_{sig}t + 2\varphi) + \frac{E_{LO}^2}{2}\cos(2\omega_{LO}t) + E_{sig}E_{LO}\cos((\omega_{sig} + \omega_{LO})t + \varphi)}_{high\ frequency\ component}$$

$$+ \underbrace{E_{sig}E_{LO}\cos((\omega_{sig} - \omega_{LO})t + \varphi)}_{beat\ component}.$$

The output has high frequency ($2\omega_{sig}$, $2\omega_{LO}$ and $\omega_{sig} + \omega_{LO}$) and constant components. In heterodyne detection, the high frequency components and usually the constant components are filtered out, leaving the intermediate (beat) frequency at $\omega_{sig} - \omega_{LO}$. The amplitude of this last component is proportional to the amplitude of the signal radiation. With appropriate signal analysis the phase of the signal can be recovered as well.

If ω_{LO} is equal to ω_{sig} then the beat component is a recovered version of the original signal, with the amplitude equal to the product of E_{sig} and E_{LO}; that is, the received signal is amplified by mixing with the local oscillator. This is the basis for a Direct conversion receiver.

Implementations

Multiplicative mixers have been implemented in a wide variety of ways. The most popular are Gilbert cell mixers, diode mixers, diode ring mixers (ring modulation) and switching mixers. Diode mixers take advantage of the non-linearity of diode devices to produce the desired multiplication in the squared term. It is a very inefficient method as most of the power output is in other unwanted terms which need filtering out. Inexpensive AM radios still use diode mixers.

Electronic mixers are usually made with transistors and/or diodes arranged in a balanced circuit or even a double-balanced circuit. These are readily manufactured by using the technology of either monolithic integrated circuits or hybrid integrated circuits. These are designed for a wide variety of frequency ranges, and they are mass-produced to tight tolerances by the hundreds of thousands, making them relatively cheap components.

These double-balanced mixers are very widely used in microwave communication systems, satellite communication systems, and ultrahigh frequency (UHF) communications transmitters and receivers, and in radar systems transmitters and receivers.

Gilbert cell mixers are just an arrangement of transistors that multiplies the two signals. The switching mixers (below) pass more power and usually insert less distortion.

switching mixers use arrays of field-effect transistors or (in older days) vacuum tubes. These are used as electronic switches, to permit the signal to go one direction, then the other. They are controlled by the signal being mixed. They are especially popular with digitally controlled radios.

Stem Mixing and Mastering

Stem-mixing is a method of mixing audio material based on creating groups of audio tracks and processing them separately prior to combining them into a final master mix. Stems are also sometimes referred to as submixes, subgroups, or busses.

Some people consider stems the same as separation mastering although others consider stems to be sub-mixes to be used with separation mastering. There is some lack of clarity with regards to what *is* a stem versus what is a separation. Semantically it seems to depend on how many separate channels of input are available for mixing and/or at which stage they are towards reducing them down a final stereo mix, with different people drawing the separation line at different places.

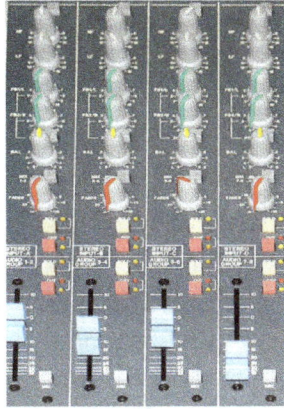

Image of Sub-group (stem) busses on a mixing console.

Image of group assign features on individual mix channels of a mixing console.
Green arrows indicate group assign buttons.

This technique originated in the 1960s with the introduction of mixing boards that were equipped with abilities to assign individual inputs to sub-group faders and then manipulate each sub-group (stem mix) independently from the others. This technique is widely used in recording studios to control, process and manipulate entire groups of instruments such as drums, strings, or backup vocals, in order to streamline and simplify the mixing process. Additionally, as each stem-bus usually has its own inserts, sends and returns, the stem-mix (sub-mix) can be processed independently through its own signal processing chain to achieve a different effect for each group of instruments. This technique is also practiced with Digital audio workstations (DAWs) in a similar way where groups of audio tracks may be processed and manipulated digitally through a separate chain of plugins.

Stem-mastering is a technique derived from stem mixing. Just like in stem-mixing, the individual audio tracks are grouped together to allow for independent control and signal processing of each stem and can be manipulated independently from each other. Even though this method is not commonly practiced by mastering studios it does have its proponents.

Stem

In audio production, a stem is a group of audio sources mixed together, usually by one person, to be dealt with downstream as one unit. A single stem may be delivered in mono, stereo, or in multiple tracks for surround sound.

In sound mixing for film, the preparation of stems is a common stratagem to facilitate the final mix. Dialogue, music and sound effects, called "D-M-E", are brought to the final mix as separate stems. Using stem mixing, the dialogue can easily be replaced by a foreign language version, the effects can easily be adapted to different mono, stereo and surround systems, and the music can be changed to fit the desired emotional response. If the music and effects stems are sent to another production facility for foreign dialogue replacement, these non-dialogue stems are called "M&E". The dialogue stem is used by itself when editing various scenes together to construct a trailer of the film; after this some music and effects are mixed in to form a cohesive sequence.

In music mixing for recordings and for live sound, stems are subgroups of similar sound sources. When a large project uses more than one person mixing, stems can facilitate the job of the final mix engineer. Such stems may consist of all of the string instruments, a full orchestra, just background vocals, only the percussion instruments, a single drum set, or any other grouping that may ease the task of the final mix. Stems prepared in this fashion may be blended together later in time, as for a recording project or for consumer listening, or they may be mixed simultaneously, as in a live sound performance with multiple elements. For instance, when Barbra Streisand toured in 2006 and 2007, the audio production crew used three people to run three mixing consoles: one to mix strings, one to mix brass, reeds and percussion, and one under main engineer Bruce Jackson's control out in the audience, containing Streisand's microphone inputs and stems from the other two consoles.

Stems may be supplied to a musician in the recording studio so that the musician can adjust a headphones monitor mix by varying the levels of other instruments and vocals relative to the musician's own input. Stems may also be delivered to the consumer so they can listen to a piece of music with a custom blend of the separate elements.

References

- Shannon Slaton (June 1, 2008). "Sound Product of the Month: Dugan Model E-1 Automatic Mixing Controller". Livedesignonline.com. Retrieved 2011-03-12

- Rumsey, Francis; McCormick, Tim (2009). Sound and Recording (6th ed.). Oxford, United Kingdom: Elsevier Inc. p. 168. ISBN 978-0-240-52163-3

- Dan Dugan (October 1989). "Application of Automatic Mixing Techniques to Audio Consoles". Audio Engineering Society. Retrieved 2011-03-12

- Wilkins, Trev (2007). Access All Areas: A Real World Guide to Gigging and Touring. Taylor & Francis US. pp. 130, 176–178, 272. ISBN 0240520440

- Dugan, Daniel W. (February 1991). "Adaptive Threshold Automatic Microphone Mixing System Becomes Public Domain". Audio Engineering Society. Retrieved 2011-03-12

- Hollyn, Norman (2009). The Film Editing Room Handbook: How to Tame the Chaos of the Editing Room (4 ed.). Peachpit Press. pp. 162, 186, 269. ISBN 0321679520

- "Dan Dugan Sound Design". National Association of Broadcasters. Archived from the original on March 21, 2012. Retrieved March 21, 2011

5

A Comprehensive Study of Audio Amplifiers

Amplifiers are devices used to increase the voltage or power of a signal. Headphone amplifier, class-t amplifier, carbon microphone, and valve audio amplifier are some examples of audio amplifiers that have been listed in this chapter. The section on audio amplifiers offers an insightful focus, keeping in mind the complex subject matter.

Amplifier

A 100 watt stereo audio amplifier used in home component audio systems in the 1970s.

An amplifier, electronic amplifier or (informally) amp is an electronic device that can increase the power of a signal (a time-varying voltage or current). An amplifier uses electric power from a power supply to increase the amplitude of a signal. The amount of amplification provided by an amplifier is measured by its gain: the ratio of output to input. An amplifier is a circuit that can give a power gain greater than one.

An amplifier can either be a separate piece of equipment or an electrical circuit contained within another device. Amplification is fundamental to modern electronics, and amplifiers are widely used in almost all electronic equipment. Amplifiers can be categorized in different ways. One is by the frequency of the electronic signal being amplified; audio amplifiers amplify signals in the audio (sound) range of less than 20 kHz, RF amplifiers amplify frequencies in the radio frequency range between 20 kHz and 300 GHz, and servo amplifiers and instrumentation amplifiers may work with very low frequencies down to direct current. A further distinction is whether the output is a linear or nonlinear representation of the input. Amplifiers can also be categorized by their physical placement in the signal chain; a preamplifier may precede other signal processing stages, for example.

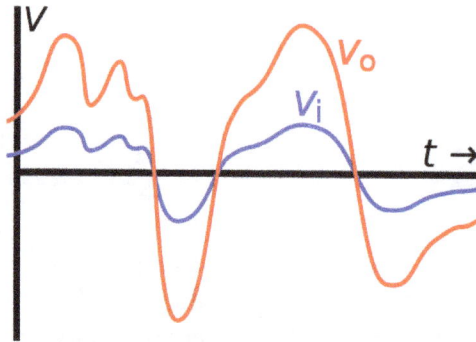

Graph of the input *(blue)* and output voltage *(red)* of an ideal linear amplifier with an arbitrary signal applied as input. *Amplification* means increasing the amplitude (voltage or current) of a time-varying signal by a given factor, as shown here. In this example the amplifier has a *voltage gain* of 3; that is at any instant

After electro-mechanical signal repeaters, such as the Shreeve repeater of 1904, the first practical and entirely electronic device that could amplify was the triode vacuum tube, invented in 1906 by Lee De Forest, which led to the first amplifiers around 1912. Vacuum tubes were used in almost all amplifiers until the 1960s–1970s when the transistor, invented in 1947, replaced them. Today, most amplifiers use transistors, but vacuum tubes continue to be used in some applications.

History

De Forest's prototype audio amplifier of 1914. This Audion (triode) vacuum tube had a voltage gain of about 5, providing a total gain of approximately 125 for this three-stage amplifier.

The development of audio communication technology in form of the telephone, first patented in 1876, created the need to increase the amplitude of electrical signals to extend the transmission of signals over increasingly long distances. In telegraphy, this problem had been solved with intermediate devices at stations that replenished the dissipated energy by operating a signal recorder and transmitter back-to-back, forming a relay, so that a local energy source at each intermediate station powered the next leg of transmission. For duplex transmission, i.e. sending and receiving in both directions, bi-directional relay repeaters were developed starting with the work of C. F. Varley for telegraphic transmission. Duplex transmission was essential for telephony and the problem was not satisfactorily solved until 1904, when H. E. Shreeve of the American Telephone and Telegraph Company improved existing attempts at constructing a telephone repeater consisting of back-to-back carbon-granule transmitter and electrodynamic receiver pairs. The Shreeve repeater was first tested on a line between Boston and Amesbury, MA, and more refined devices remained in service for some time.

After the turn of the century it was found that negative resistance mercury lamps could amplify, and were also tried in repeaters. The concurrent development of thermionic valves starting ca. 1902, provided an entirely electronic method of amplifying signals. The first practical version of such devices was the Audion triode, invented in 1906 by Lee De Forest, which led to the first amplifiers around 1912. In analogy to previous types of relays in telegraphy and telephony, the amplifying vacuum tube was first called an *electron relay*. In the first extensive commercial use of the vacuum tube, such repeaters powered the first transcontinental telephone line for commercial service in 1915.

The terms *amplifier* and *amplification*, derived from the Latin *amplificare*, (*to enlarge or expand*), were first used for this new capability around 1915 when triodes became widespread.

The amplifying vacuum tube revolutionized electrical technology, creating the new field of electronics, the technology of active electrical devices. It made possible long distance telephone lines, public address systems, radio broadcasting, talking motion pictures, practical audio recording, radar, television, and the first computers. For 50 years virtually all consumer electronic devices used vacuum tubes. Early tube amplifiers often had positive feedback (regeneration), which could increase gain but also make the amplifier unstable and prone to oscillation. Much of the mathematical theory of amplifiers was developed at Bell Telephone Laboratories during the 1920s to 1940s. Distortion levels in early amplifiers were high, usually around 5%, until 1934, when Harold Black developed negative feedback; this allowed the distortion levels to be greatly reduced, at the cost of lower gain. Other advances in the theory of amplification were made by Harry Nyquist and Hendrik Wade Bode.

The Shreeve mechanical repeater and the vacuum tube were the only amplifying devices, other than specialized power devices such as the magnetic amplifier and amplidyne, for 40 years. Power control circuitry used magnetic amplifiers until the latter half of the twentieth century when power semiconductor devices became more economical, with higher operating speeds. Shreeve repeaters were used as adjustable amplifiers in telephone subscriber sets for the hearing impaired until the transistor provided smaller and higher quality amplifiers in the 1950s. The replacement of bulky electron tubes with transistors during the 1960s and 1970s created another revolution in electronics, making possible a large class of portable electronic devices, such as the transistor radio developed in 1954. Today, use of vacuum tubes is limited for some high power applications, such as radio transmitters.

Beginning in the 1970s, more and more transistors were connected on a single chip thereby creating higher scales of integration (small-scale, medium-scale, large-scale, etc.) in integrated circuits. Many amplifiers commercially available today are based on integrated circuits.

For special purposes, other active elements have been used. For example, in the early days of the satellite communication, parametric amplifiers were used. The core circuit was a diode whose capacitance was changed by an RF signal created locally. Under certain conditions, this RF signal provided energy that was modulated by the extremely weak satellite signal received at the earth station.

Advances in digital electronics since the late 20th century provided new alternatives to the traditional linear-gain amplifiers by using digital switching to vary the pulse-shape of fixed amplitude signals, resulting in devices such as the Class-D amplifier.

Ideal

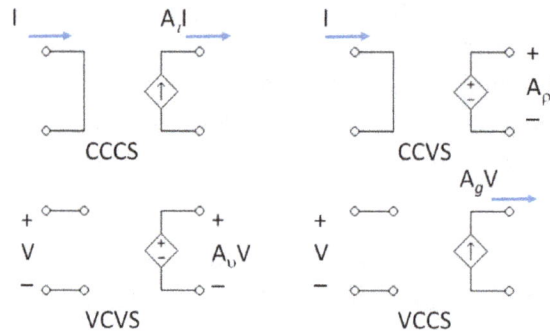

The four types of dependent source—control variable on left, output variable on right

In principle, an amplifier is an electrical two-port network that provides an output signal that is a replica of the input signal, but increased in magnitude.

Electronic amplifiers use one variable presented as either a current and voltage. Either current or voltage can be used as input and either as output, leading to four types of amplifiers. In idealized form they are represented by each of the four types of dependent source used in linear analysis, as shown in the figure, namely:

Input	Output	Dependent source	Amplifier type	Gain units
I	I	Current controlled current source, CCCS	Current amplifier	Unitless
I	V	Current controlled voltage source, CCVS	Transresistance amplifier	Ohm
V	I	Voltage controlled current source, VCCS	Transconductance amplifier	Siemens
V	V	Voltage controlled voltage source, VCVS	Voltage amplifier	Unitless

Each type of amplifier in its ideal form has an ideal input and output resistance that is the same as that of the corresponding dependent source:

Amplifier type	Dependent source	Input impedance	Output impedance
Current	CCCS	0	∞
Transresistance	CCVS	0	0
Transconductance	VCCS	∞	∞
Voltage	VCVS	∞	0

In practice the ideal impedances are not possible to achieve. For any particular circuit, a small-signal analysis is often used to find the actual impedance. A small-signal AC test current I_x is applied to the input or output node, all external sources are set to AC zero, and the corresponding alternating voltage V_x across the test current source determines the impedance seen at that node as $R = V_x / I_x$.

Amplifiers designed to attach to a transmission line at input and output, especially RF amplifiers, do not fit into this classification approach. Rather than dealing with voltage or current individually, they ideally couple with an input or output impedance matched to the transmission line impedance, that is, match *ratios* of voltage to current. Many real RF amplifiers come close to this ideal. Although, for a given appropriate source and load impedance, RF amplifiers can be characterized as amplifying voltage or current, they fundamentally are amplifying power.

Properties

Amplifier properties are given by parameters that include:

- Gain, the ratio between the magnitude of output and input signals

- Bandwidth, the width of the useful frequency range

- Efficiency, the ratio between the power of the output and total power consumption

- Linearity, the extent to which the proportion between input and output amplitude is the same for high amplitude and low amplitude input

- Noise, a measure of undesired noise mixed into the output

- Output dynamic range, the ratio of the largest and the smallest useful output levels

- Slew rate, the maximum rate of change of the output

- Rise time, settling time, ringing and overshoot that characterize the step response

- Stability, the ability to avoid self-oscillation

Amplifiers are described according to the properties of their inputs, their outputs, and how they relate. All amplifiers have gain, a multiplication factor that relates the magnitude of some property of the output signal to a property of the input signal. The gain may be specified as the ratio of output voltage to input voltage (voltage gain), output power to input power (power gain), or some combination of current, voltage, and power. In many cases the property of the output that varies is dependent on the same property of the input, making the gain unitless (though often expressed in decibels (dB)).

Most amplifiers are designed to be linear. That is, they provide constant gain for any normal input level and output signal. If an amplifier's gain is not linear, the output signal can become distorted. There are, however, cases where variable gain is useful. Certain signal processing applications use exponential gain amplifiers.

Amplifiers are usually designed to function well in a specific application, for example: radio and television transmitters and receivers, high-fidelity ("hi-fi") stereo equipment, microcomputers and other digital equipment, and guitar and other instrument amplifiers. Every amplifier includes at least one active device, such as a vacuum tube or transistor.

Negative Feedback

Negative feedback feeds the difference of the input and part of the output back to the input in a way that cancels out part of the input. The main effect is to reduce the overall gain of the system.

However, the unwanted signals introduced by the amplifier are also fed back. Since they are not part of the original input, they are added to the input in opposite phase, subtracting them from the input. In this way, negative feedback acts as a technique to reduce errors (at the expense of gain). Large amounts of negative can reduce errors to the point that the response of the amplifier itself becomes almost irrelevant as long as it has a large gain, and the output performance of the system (the "closed loop performance") is defined entirely by the components in the feedback loop.

Careful design of each stage of an open-loop (non-feedback) amplifier can achieve about 1% distortion for audio-frequency signals. With negative feedback, 0.001% is typical. Noise, even crossover distortion, can be practically eliminated. Negative feedback also compensates for changing temperatures, and degrading or nonlinear, components in the gain stage, but any change or nonlinearity in the components in the feedback loop will affect the output. Indeed, the ability of the feedback loop to define the output is used to make active filter circuits. The concept of feedback is used in operational amplifiers to precisely define gain, bandwidth, and other parameters entirely based on the components in the feedback loop.

Negative feedback can be applied at each stage of an amplifier to stabilize the operating point of active devices against minor changes in power-supply voltage or device characteristics.

Some feedback, positive or negative, is unavoidable and often undesirable—introduced, for example, by parasitic elements, such as inherent capacitance between input and output of devices such as transistors, and capacitive coupling of external wiring. Excessive frequency-dependent positive feedback can produce parasitic oscillation and turn an amplifier into an oscillator.

Categories

Active Devices

All amplifiers include some form of active device: this is the device that does the actual amplification. The active device can be a vacuum tube, discrete solid state component, such as a single transistor, or part of an integrated circuit, as in an op-amp).

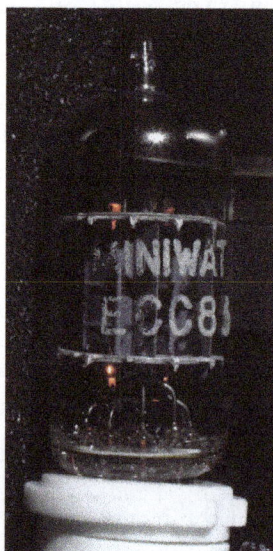

An ECC83 tube glowing inside a preamp.

Transistor amplifiers (or solid state amplifiers) are the most common type of amplifier in use today. A transistor is used as the active element. The gain of the amplifier is determined by the properties of the transistor itself as well as the circuit it is contained within.

Common active devices in transistor amplifiers include bipolar junction transistors (BJTs) and metal oxide semiconductor field-effect transistors (MOSFETs).

Applications are numerous, some common examples are audio amplifiers in a home stereo or public address system, RF high power generation for semiconductor equipment, to RF and microwave applications such as radio transmitters.

Transistor-based amplification can be realized using various configurations: for example a bipolar junction transistor can realize common base, common collector or common emitter amplification; a MOSFET can realize common gate, common source or common drain amplification. Each configuration has different characteristics.

Vacuum-tube amplifiers (also known as tube amplifiers or valve amplifiers) use a vacuum tube as the active device. While semiconductor amplifiers have largely displaced valve amplifiers for low-power applications, valve amplifiers can be much more cost effective in high power applications such as radar, countermeasures equipment, and communications equipment. Many microwave amplifiers are specially designed valve amplifiers, such as the klystron, gyrotron, traveling wave tube, and crossed-field amplifier, and these microwave valves provide much greater single-device power output at microwave frequencies than solid-state devices. Vacuum tubes remain in use in some high end audio equipment, as well as in musical instrument amplifiers, due to a preference for "tube sound".

Magnetic amplifiers are devices somewhat similar to a transformer where one winding is used to control the saturation of a magnetic core and hence alter the impedance of the other winding.

They have largely fallen out of use due to development in semiconductor amplifiers but are still useful in HVDC control, and in nuclear power control circuitry due to not being affected by radioactivity.

Negative resistances can be used as amplifiers, such as the tunnel diode amplifier.

Power Amplifiers

Power amplifier by Skyworks Solutions in a Smartphone.

A power amplifier is an amplifier designed primarily to increase the power available to a load. In practice, amplifier power gain depends on the source and load impedances, as well as the inherent voltage and current gain. A radio frequency (RF) amplifier design typically optimizes impedances for power transfer, while audio and instrumentation amplifier designs normally optimize input and output impedance for least loading and highest signal integrity. An amplifier that is said to have a gain of 20 dB might have a voltage gain of 20 dB and an available power gain of much more than 20 dB (power ratio of 100)—yet actually deliver a much lower power gain if, for example, the input is from a 600 Ω microphone and the output connects to a 47 kΩ input socket for a power amplifier. In general the power amplifier is the last 'amplifier' or actual circuit in a signal chain (the output stage) and is the amplifier stage that requires attention to power efficiency. Efficiency considerations lead to the various classes of power amplifier based on the biasing of the output transistors or tubes.

Audio power amplifiers are typically used to drive loudspeakers. They will often have two output channels and deliver equal power to each. An RF power amplifier is found in radio transmitter final stages. A Servo motor controller: amplifies a control voltage to adjust the speed of a motor, or the position of a motorized system.

Operational Amplifiers (Op-amps)

An LM741 general purpose op-amp

An operational amplifier is an amplifier circuit which typically has very high open loop gain and differential inputs. Op amps have become very widely used as standardized "gain blocks" in circuits due to their versatility; their gain, bandwidth and other characteristics can be controlled by feedback through an external circuit. Though the term today commonly applies to integrated circuits, the original operational amplifier design used valves, and later designs used discrete transistor circuits.

A fully differential amplifier is similar to the operational amplifier, but also has differential outputs. These are usually constructed using BJTs or FETs.

Distributed Amplifiers

These use balanced transmission lines to separate individual single stage amplifiers, the outputs of which are summed by the same transmission line. The transmission line is a balanced type with the input at one end and on one side only of the balanced transmission line and the output at the opposite end is also the opposite side of the balanced transmission line. The gain of each stage adds linearly to the output rather than multiplies one on the other as in a cascade configuration. This allows a higher bandwidth to be achieved than could otherwise be realised even with the same gain stage elements.

Switched Mode Amplifiers

These nonlinear amplifiers have much higher efficiencies than linear amps, and are used where the power saving justifies the extra complexity. Class-D amplifiers are the main example of this type of amplification.

Applications

Video Amplifiers

Video amplifiers are designed to process video signals and have varying bandwidths depending on whether the video signal is for SDTV, EDTV, HDTV 720p or 1080i/p etc.. The specification of the bandwidth itself depends on what kind of filter is used—and at which point (−1 dB or −3 dB for example) the bandwidth is measured. Certain requirements for step response and overshoot are necessary for an acceptable TV image.

Microwave Amplifiers

Traveling wave tube amplifiers (TWTAs) are used for high power amplification at low microwave frequencies. They typically can amplify across a broad spectrum of frequencies; however, they are usually not as tunable as klystrons.

Klystrons are specialized linear-beam vacuum-devices, designed to provide high power, widely tunable amplification of millimetre and sub-millimetre waves. Klystrons are designed for large scale operations and despite having a narrower bandwidth than TWTAs, they have the advantage of coherently amplifying a reference signal so its output may be precisely controlled in amplitude, frequency and phase.

Solid-state devices are used such as GaAs FETs, IMPATT diodes, and others, especially at lower microwave frequencies and power levels on the order of watts.

The maser is a non-electronic microwave amplifier.

Musical Instrument Amplifiers

Instrument amplifiers are a range of audio power amplifiers used to increase the sound level of musical instruments, for example guitars, during performances.

Classification of Amplifier Stages and Systems

Common Terminal

One set of classifications for amplifiers is based on which device terminal is common to both the input and the output circuit. In the case of bipolar junction transistors, the three classes are common emitter, common base, and common collector. For field-effect transistors, the corresponding configurations are common source, common gate, and common drain; for vacuum tubes, common cathode, common grid, and common plate.

The common emitter (or common source, common cathode, etc.) is most often configured to provide amplification of a voltage applied between base and emitter, and the output signal taken between collector and emitter is inverted, relative to the input. The common collector arrangement applies the input voltage between base and collector, and to take the output voltage between emitter and collector. This causes negative feedback, and the output voltage tends to follow the input voltage. This arrangement is also used as the input presents a high impedance and does not load the signal source, though the voltage amplification is less than one. The common-collector circuit is, therefore, better known as an emitter follower, source follower, or cathode follower.

Unilateral or Bilateral

An amplifier whose output exhibits no feedback to its input side is described as 'unilateral'. The input impedance of a unilateral amplifier is independent of load, and output impedance is independent of signal source impedance.

An amplifier that uses feedback to connect part of the output back to the input is a *bilateral* amplifier. Bilateral amplifier input impedance depends on the load, and output impedance on the signal source impedance. All amplifiers are bilateral to some degree; however they may often be modeled as unilateral under operating conditions where feedback is small enough to neglect for most purposes, simplifying analysis.

Inverting or Non-inverting

Another way to classify amplifiers is by the phase relationship of the input signal to the output signal. An 'inverting' amplifier produces an output 180 degrees out of phase with the input signal (that is, a polarity inversion or mirror image of the input as seen on an oscilloscope). A 'non-inverting' amplifier maintains the phase of the input signal waveforms. An emitter follower is a type of non-inverting amplifier, indicating that the signal at the emitter of a transistor is following (that is, matching with unity gain but perhaps an offset) the input signal. Voltage follower is also non inverting type of amplifier having unity gain.

This description can apply to a single stage of an amplifier, or to a complete amplifier system.

Function

Other amplifiers may be classified by their function or output characteristics. These functional descriptions usually apply to complete amplifier systems or sub-systems and rarely to individual stages.

- A servo amplifier indicates an integrated feedback loop to actively control the output at some desired level. A DC servo indicates use at frequencies down to DC levels, where the rapid fluctuations of an audio or RF signal do not occur. These are often used in mechanical actuators, or devices such as DC motors that must maintain a constant speed or torque. An AC servo amp. can do this for some AC motors.

- A linear amplifier responds to different frequency components independently, and does not generate harmonic distortion or intermodulation distortion. No amplifier can provide *perfect* linearity (even the most linear amplifier has some nonlinearities, since the amplifying devices—transistors or vacuum tubes—follow nonlinear power laws such as square-laws and rely on circuitry techniques to reduce those effects).

- A nonlinear amplifier generates significant distortion and so changes the harmonic content; there are situations where this is useful. Amplifier circuits intentionally providing a non-linear transfer function include:

  o a device like a silicon controlled rectifier or a transistor used as a switch may be employed to turn either fully *on* or *off* a load such as a lamp based on a threshold in a continuously variable input.

  o a non-linear amplifier in an analog computer or true RMS converter for example can provide a special transfer function, such as logarithmic or square-law.

  o a Class C RF amplifier may be chosen because it can be very efficient—but is non-linear. Following such an amplifier with a so-called *tank* tuned circuit can reduce unwanted harmonics (distortion) sufficiently to make it useful in transmitters, or some desired harmonic may be selected by setting the resonant frequency of the tuned circuit to a higher frequency rather than fundamental frequency in frequency multiplier circuits.

  o Automatic gain control circuits require an amplifier's gain be controlled by the time-averaged amplitude so that the output amplitude varies little when weak stations are being received. The non-linearities are assumed arranged so the relatively small signal amplitude suffers from little distortion (cross-channel interference or intermodulation) yet is still modulated by the relatively large gain-control DC voltage.

  o AM detector circuits that use amplification such as anode-bend detectors, precision rectifiers and infinite impedance detectors (so excluding *unamplified* detectors such as cat's-whisker detectors), as well as peak detector circuits, rely on changes in amplification based on the signal's instantaneous amplitude to derive a direct current from an alternating current input.

  o Operational amplifier comparator and detector circuits.

- A wideband amplifier has a precise amplification factor over a wide frequency range, and is often used to boost signals for relay in communications systems. A narrowband amp amplifies a specific narrow range of frequencies, to the exclusion of other frequencies.

- An RF amplifier amplifies signals in the radio frequency range of the electromagnetic spectrum, and is often used to increase the sensitivity of a receiver or the output power of a transmitter.

- An audio amplifier amplifies audio frequencies. This category subdivides into small signal amplification, and power amps that are optimised to driving speakers, sometimes with multiple amps grouped together as separate or bridgeable channels to accommodate different audio reproduction requirements. Frequently used terms within audio amplifiers include:

o   Preamplifier (preamp.), which may include a phono preamp with RIAA equaliza-
    tion, or tape head preamps with CCIR equalisation filters. They may include filters
    or tone control circuitry.

o   Power amplifier (normally drives loudspeakers), headphone amplifiers, and public
    address amplifiers.

o   Stereo amplifiers imply two channels of output (left and right), though the term
    simply means "solid" sound (referring to three-dimensional)—so quadraphonic
    stereo was used for amplifiers with four channels. 5.1 and 7.1 systems refer to Home
    theatre systems with 5 or 7 normal spatial channels, plus a subwoofer channel.

• Buffer amplifiers, which may include emitter followers, provide a high impedance input
  for a device (perhaps another amplifier, or perhaps an energy-hungry load such as lights)
  that would otherwise draw too much current from the source. Line drivers are a type of
  buffer that feeds long or interference-prone interconnect cables, possibly with differential
  outputs through twisted pair cables.

Interstage Coupling Method

Amplifiers are sometimes classified by the coupling method of the signal at the input, output, or
between stages. Different types of these include:

Resistive-capacitive (RC) coupled amplifier, using a network of resistors and capacitors

By design these amplifiers cannot amplify DC signals as the capacitors block the DC component of the input signal. RC-coupled amplifiers were used very often in circuits with
vacuum tubes or discrete transistors. In the days of the integrated circuit a few more transistors on a chip are much cheaper and smaller than a capacitor.

Inductive-capacitive (LC) coupled amplifier, using a network of inductors and capacitors

This kind of amplifier is most often used in selective radio-frequency circuits.

Transformer coupled amplifier, using a transformer to match impedances or to decouple parts of
the circuits

Quite often LC-coupled and transformer-coupled amplifiers cannot be distinguished as a
transformer is some kind of inductor.

Direct coupled amplifier, using no impedance and bias matching components

This class of amplifier was very uncommon in the vacuum tube days when the anode (output) voltage was at greater than several hundred volts and the grid (input) voltage at a few
volts minus. So they were only used if the gain was specified down to DC (e.g., in an oscilloscope). In the context of modern electronics developers are encouraged to use directly coupled amplifiers whenever possible. In FET and CMOS technologies direct coupling
is dominant since gates of MOSFETs theoretically pass no current through themselves.
Therefore, DC component of the input signals is automatically filtered.

Frequency Range

Depending on the frequency range and other properties amplifiers are designed according to different principles.

Frequency ranges down to DC are only used when this property is needed. Amplifiers for direct current signals are vulnerable to minor variations in the properties of components with time. Special methods, such as chopper stabilized amplifiers are used to prevent objectionable drift in the amplifier's properties for DC. "DC-blocking" capacitors can be added to remove DC and sub-sonic frequencies from audio amplifiers.

Depending on the frequency range specified different design principles must be used. Up to the MHz range only "discrete" properties need be considered; e.g., a terminal has an input impedance.

As soon as any connection within the circuit gets longer than perhaps 1% of the wavelength of the highest specified frequency (e.g., at 100 MHz the wavelength is 3 m, so the critical connection length is approx. 3 cm) design properties radically change. For example, a specified length and width of a PCB trace can be used as a selective or impedance-matching entity. Above a few hundred MHz, it gets difficult to use discrete elements, especially inductors. In most cases, PCB traces of very closely defined shapes are used instead (stripline techniques).

The frequency range handled by an amplifier might be specified in terms of bandwidth (normally implying a response that is 3 dB down when the frequency reaches the specified bandwidth), or by specifying a frequency response that is within a certain number of decibels between a lower and an upper frequency (e.g. "20 Hz to 20 kHz plus or minus 1 dB").

Power Amplifier Classes

Power amplifier circuits (output stages) are classified as A, B, AB and C for analog designs—and class D and E for switching designs. The power amplifier classes are based on the proportion of each input cycle (conduction angle) during which an amplifying device passes current. The image of the conduction angle derives from amplifying a sinusoidal signal. If the device is always on, the conducting angle is 360°. If it is on for only half of each cycle, the angle is 180°. The angle of flow is closely related to the amplifier power efficiency.

Example Amplifier Circuit

The practical amplifier circuit to the right could be the basis for a moderate-power audio amplifier. It features a typical (though substantially simplified) design as found in modern amplifiers, with a class-AB push–pull output stage, and uses some overall negative feedback. Bipolar transistors are shown, but this design would also be realizable with FETs or valves.

The input signal is coupled through capacitor C1 to the base of transistor Q1. The capacitor allows the AC signal to pass, but blocks the DC bias voltage established by resistors R1 and R2 so that any preceding circuit is not affected by it. Q1 and Q2 form a differential amplifier (an amplifier that multiplies the difference between two inputs by some constant), in an arrangement known as a long-tailed pair. This arrangement is used to conveniently allow the use of negative feedback, which is fed from the output to Q2 via R7 and R8.

A practical amplifier circuit

The negative feedback into the difference amplifier allows the amplifier to compare the input to the actual output. The amplified signal from Q1 is directly fed to the second stage, Q3, which is a common emitter stage that provides further amplification of the signal and the DC bias for the output stages, Q4 and Q5. R6 provides the load for Q3 (a better design would probably use some form of active load here, such as a constant-current sink). So far, all of the amplifier is operating in class A. The output pair are arranged in class-AB push–pull, also called a complementary pair. They provide the majority of the current amplification (while consuming low quiescent current) and directly drive the load, connected via DC-blocking capacitor C2. The diodes D1 and D2 provide a small amount of constant voltage bias for the output pair, just biasing them into the conducting state so that crossover distortion is minimized. That is, the diodes push the output stage firmly into class-AB mode (assuming that the base-emitter drop of the output transistors is reduced by heat dissipation).

This design is simple, but a good basis for a practical design because it automatically stabilises its operating point, since feedback internally operates from DC up through the audio range and beyond. Further circuit elements would probably be found in a real design that would roll-off the frequency response above the needed range to prevent the possibility of unwanted oscillation. Also, the use of fixed diode bias as shown here can cause problems if the diodes are not both electrically and thermally matched to the output transistors – if the output transistors turn on too much, they can easily overheat and destroy themselves, as the full current from the power supply is not limited at this stage.

A common solution to help stabilise the output devices is to include some emitter resistors, typically one ohm or so. Calculating the values of the circuit's resistors and capacitors is done based on the components employed and the intended use of the amp.

Notes on Implementation

Any real amplifier is an imperfect realization of an ideal amplifier. An important limitation of a real amplifier is that the output it generates is ultimately limited by the power available from the power supply. An amplifier saturates and clips the output if the input signal becomes too large for the amplifier to reproduce or exceeds operational limits for the device. The power supply may influence the output, so must be considered in the design. The power output from an amplifier cannot exceed its input power.

The amplifier circuit has an "open loop" performance. This is described by various parameters (gain, slew rate, output impedance, distortion, bandwidth, signal to noise ratio, etc.). Many modern amplifiers use negative feedback techniques to hold the gain at the desired value and reduce distortion. Negative loop feedback has the intended effect of electrically damping loudspeaker motion, thereby damping the mechanical dynamic performance of the loudspeaker.

When assessing rated amplifier power output, it is useful to consider the applied load, the signal type (e.g., speech or music), required power output duration (i.e., short-time or continuous), and required dynamic range (e.g., recorded or live audio). In high-powered audio applications that require long cables to the load (e.g., cinemas and shopping centres) it may be more efficient to connect to the load at line output voltage, with matching transformers at source and loads. This avoids long runs of heavy speaker cables.

To prevent instability or overheating requires care to ensure solid state amplifiers are adequately loaded. Most have a rated minimum load impedance.

All amplifiers generate heat through electrical losses. The amplifier must dissipate this heat via convection or forced air cooling. Heat can damage or reduce electronic component service life. Designers and installers must also consider heating effects on adjacent equipment.

Different power supply types result in many different methods of bias. Bias is a technique by which active devices are set to operate in a particular region, or by which the DC component of the output signal is set to the midpoint between the maximum voltages available from the power supply. Most amplifiers use several devices at each stage; they are typically matched in specifications except for polarity. Matched inverted polarity devices are called complementary pairs. Class-A amplifiers generally use only one device, unless the power supply is set to provide both positive and negative voltages, in which case a dual device symmetrical design may be used. Class-C amplifiers, by definition, use a single polarity supply.

Amplifiers often have multiple stages in cascade to increase gain. Each stage of these designs may be a different type of amp to suit the needs of that stage. For instance, the first stage might be a class-A stage, feeding a class-AB push–pull second stage, which then drives a class-G final output stage, taking advantage of the strengths of each type, while minimizing their weaknesses.

Preamplifier

An example of a typical high-end stereo preamplifier.

A preamplifier (preamp or "pre") is an electronic amplifier that converts a weak electrical signal into an output signal strong enough to be noise-tolerant and strong enough for further processing, or for sending to a power amplifier and a loudspeaker. Without this, the final signal would be noisy or distorted. They are typically used to amplify signals from analog sensors such as microphones and pickups. Because of this, the preamplifier is often placed close to the sensor to reduce the effects of noise and interference.

Description

An ideal preamp will be linear (have a constant gain through its operating range), have high input impedance (requiring only a minimal amount of current to sense the input signal) and a low output impedance (when current is drawn from the output there is minimal change in the output voltage). It is used to boost the signal strength to drive the cable to the main instrument without significantly degrading the signal-to-noise ratio (SNR). The noise performance of a preamplifier is critical. According to Friis's formula, when the gain of the preamplifier is high, the SNR of the final signal is determined by the SNR of the input signal and the noise figure of the preamplifier.

Three basic types of preamplifiers are available:

- current-sensitive preamplifier
- parasitic-capacitance preamplifier
- charge-sensitive preamplifier.

Audio Systems

In an audio system, they are typically used to amplify signals from analog sensors to line level. The second amplifier is typically a power amplifier (power amp). The preamplifier provides voltage gain (e.g., from 10 mV to 1 V) but no significant current gain. The power amplifier provides the higher current necessary to drive loudspeakers. For these systems some common sensors are microphones, instrument pickups, and phonographs. Preamplifiers are often integrated into the audio inputs on mixing consoles, DJ mixers, and sound cards. They can also be stand-alone devices.

Misnomers

In a home audio system, the term 'preamplifier' may sometimes be used to describe equipment which merely switches between different line level sources and applies a volume control, so that no actual pre-amplification may be involved.

Examples

- The integrated pre-amplifier in a foil electret microphone.
- The first stages of an instrument amplifier, which is then sent to the power amplifier. With instrument amplifiers, the preamp is often designed to produce overdrive or distortion effects.
- A stand-alone unit for use in live music and recording studio applications.
- As part of a stand-alone channel strip or channel strip built into an audio mixing desk.

- A masthead amplifier used with television receiver antenna or a satellite receiver dish.

- The circuit inside of a hard drive connected to the magnetic heads or the circuit inside of CD/DVD drive which connects to the photodiodes.

- A switched capacitor circuit used to null the effects of mismatch offset in most CMOS comparator-based flash analog-to-digital converters

Audio Power Amplifier

An audio power amplifier (or power amp) is an electronic amplifier that strengthens low-power, inaudible electronic audio signals such as the signal from radio receiver or electric guitar pickup to a level that is strong enough for driving (or powering) loudspeakers or headphones. This includes both amplifiers used in home audio systems and musical instrument amplifiers like guitar amplifiers. Power amplifiers make the signal—whether it is recorded music, a live speech, live singing, an electric guitar or the mixed audio of an entire band through a sound reinforcement system—audible to listeners. It is the final electronic stage in a typical audio playback chain before the signal is sent to the loudspeakers and speaker enclosures.

Audio stereo power amplifier made by Unitra

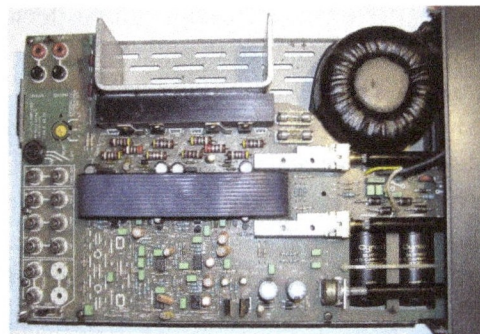

The inside of a Mission Cyrus 1 Hi Fi integrated audio amplifier (1984)

The preceding stages in such a chain are low power audio amplifiers which perform tasks like pre-amplification of the signal (this is particularly associated with record turntable signals, microphone signals and electric instrument signals from pickups, such as the electric guitar and electric bass), equalization (e.g., adjusting the bass and treble), tone controls, mixing different input signals or adding electronic effects such as reverb. The inputs can also be any number of audio sources like record players, CD players, digital audio players and cassette players. Most audio power amplifiers require these low-level inputs, which are line level.

While the input signal to an audio power amplifier, such as the signal from an electric guitar, may measure only a few hundred microwatts, its output may be a few watts for small consumer electronics devices, such as clock radios, tens or hundreds of watts for a home stereo system, several thousand watts for a nightclub's sound system or tens of thousands of watts for a large rock concert sound reinforcement system. While power amplifiers are available in standalone units, typically aimed at the hi-fi audiophile market (a niche market) of audio enthusiasts and sound reinforcement system professionals, most consumer electronics sound products, such as clock radios, boom boxes and televisions have relatively small power amplifiers that are integrated inside the chassis of the main product.

History

Three rack-mounted audio power amplifiers used in a sound reinforcement system.

The audio amplifier was invented in 1909 by Lee De Forest when he invented the triode vacuum tube (or "valve" in British English). The triode was a three terminal device with a control grid that can modulate the flow of electrons from the filament to the plate. The triode vacuum amplifier was used to make the first AM radio. Early audio power amplifiers were based on vacuum tubes and some of these achieved notably high audio quality (e.g., the Williamson amplifier of 1947-9). Audio power amplifiers based on transistors became practical with the wide availability of inexpensive transistors in the late 1960s. Since the 1970s, most modern audio amplifiers are based on solid state devices (transistors such as BJTs, FETs and MOSFETs). Transistor-based amplifiers are lighter in weight, more reliable and require less maintenance than tube amplifiers. In the 2010s, there are still audio enthusiasts, musicians (particularly electric guitarists, electric bassists, Hammond organ players and Fender Rhodes electric piano players, among others), audio engineers and music producers who prefer tube-based amplifiers, and what is perceived as a "warmer" tube sound.

Design Parameters

Key design parameters for audio power amplifiers are frequency response, gain, noise, and distortion. These are interdependent; increasing gain often leads to undesirable increases in noise and distortion. While negative feedback actually reduces the gain, it also reduces distortion. Most audio amplifiers are linear amplifiers operating in class AB. Until the 1970s, most amplifiers were tube amplifiers which used vacuum tubes. During the 1970s, tube amps were increasingly replaced with transistor-based amplifiers, which were lighter in weight, more reliable, and lower maintenance. Nevertheless, there are niche markets of consumers who continue to use tube amplifiers and tube preamplifiers in the 2010s, such as with home hi-fi enthusiasts, audio engineers and music producers (who use tube preamplifiers in studio recordings to "warm up" microphone signals) and electric guitarists, electric bassists and Hammond organ players, of whom a minority continue to use tube preamps, tube power amps and tube effects units. While hi-fi enthusiasts and audio engineers doing live sound or monitoring tracks in the studio typically seek out amplifiers with the lowest distortion, electric instrument players in genres such as blues, rock music and heavy metal music, among others, use tube amplifiers because they like the natural overdrive that tube amps produce when pushed hard. In the 2000s, the Class-D amplifier, which is much more efficient than Class AB amplifiers, is widely used in consumer electronics audio products, bass amplifiers and sound reinforcement system gear, as Class D amplifiers are much lighter in weight and produce almost no heat.

Filters and Preamplifiers

Since modern digital devices, including CD and DVD players, radio receivers and tape decks already provide a "flat" signal at line level, the preamp is not needed other than as a volume control and source selector. One alternative to a separate preamp is to simply use passive volume and switching controls, sometimes integrated into a power amplifier to form an *integrated amplifier*.

Power Output Stages

The final stage of amplification, after preamplifiers, is the output stage, where the highest demands are placed on the transistors or tubes. For this reason, the design choices made around the output device (for single-ended output stages, such as in single-ended triode amplifiers) or devices (for push-pull output stages), such as the Class of operation of the output devices is often taken as the description of the whole power amplifier. For example, a Class B amplifier will probably have just the high power output devices operating cut off for half of each cycle, while the other devices (such as differential amplifiuer, voltage amplifier and possibly even driver transistors) operate in Class A. In a transformerless output stage, the devices are essentially in series with the power supply and output load (such as a loudspeaker), possibly via some large capacitor and/or small resistances.

Further Developments

For some years following the introduction of solid state amplifiers, their perceived sound did not have the excellent audio quality of the best valve amplifiers. This led audiophiles to believe that "tube sound" or valve sound had an intrinsic quality due to the vacuum tube technology itself. In 1970, Matti Otala published a paper on the origin of a previously unobserved form of distortion: transient intermodulation distortion (TIM), later also called slew-induced distortion (SID) by others. TIM distortion was found to occur during very rapid increases in amplifier output voltage.

TIM did not appear at steady state sine tone measurements, helping to hide it from design engineers prior to 1970. Problems with TIM distortion stem from reduced open loop frequency response of solid state amplifiers. Further works of Otala and other authors found the solution for TIM distortion, including increasing slew rate, decreasing preamp frequency bandwidth, and the insertion of a lag compensation circuit in the input stage of the amplifier. In high quality modern amplifiers the open loop response is at least 20 kHz, canceling TIM distortion.

The next step in advanced design was the Baxandall Theorem, created by Peter Baxandall in England. This theorem introduced the concept of comparing the ratio between the input distortion and the output distortion of an amplifier. This new idea helped audio design engineers to better evaluate the distortion processes within an amplifier.

Applications

Important applications include public address systems, theatrical and concert sound reinforcement systems, and domestic systems such as a stereo or home-theatre system. Instrument amplifiers including guitar amplifiers and electric keyboard amplifiers also use audio power amplifiers. In some cases, the power amplifier for an instrument amplifier is integrated into a single amplifier "head" which contains a preamplifier, tone controls, and electronic effects. These components may be mounted in a wooden speaker cabinet to create a "combo amplifier". Musicians with unique performance needs and/or a need for very powerful amplification may create a custom setup with separate rackmount preamplifiers, equalizers, and a power amplifier mounted in a 19" road case.

Power amplifiers are available in standalone units, which are used by hi-fi audio enthusiasts and designers of public address systems (PA systems) and sound reinforcement systems. A hi-fi user of power amplifiers may have a stereo power amplifier to drive left and right speakers and a "monoblock" single channel power amplifier to drive a subwoofer. The number of power amplifiers used in a sound reinforcement setting depends on the size of the venue. A small coffeehouse may have a single power amp driving two PA speakers. A nightclub may have several power amps for the main speakers, one or more power amps for the monitor speakers (pointing towards the band) and an additional power amp for the subwoofer. A stadium concert may have a large number of power amps mounted in racks. Most consumer electronics sound products, such as TVs, boom boxes, home cinema sound systems, Casio and Yamaha electronic keyboards, "combo" guitar amps and car stereos have power amplifiers integrated inside the chassis of the main product.

Carbon Microphone

The carbon microphone, also known as carbon button microphone, button microphone, or carbon transmitter, is a type of microphone, a transducer that converts sound to an electrical audio signal. It consists of two metal plates separated by granules of carbon. One plate is very thin and faces toward the speaking person, acting as a diaphragm. Sound waves striking the diaphragm cause it to vibrate, exerting a varying pressure on the granules, which in turn changes the electrical resistance between the plates. Higher pressure lowers the resistance as the granules are pushed closer together. A steady direct current is passed between the plates through the granules. The varying resistance results in a modulation of the current, creating a varying electric current that reproduces the varying pressure of the sound wave. In telephony, this undulating current is directly passed through the telephone wires to the central office. In public address systems or recording devices it is amplified by an audio amplifier. The frequency response of the carbon microphone, however, is limited to a narrow range, and the device produces significant electrical noise.

Carbon microphone from Western Electric telephone handset, around 1976.

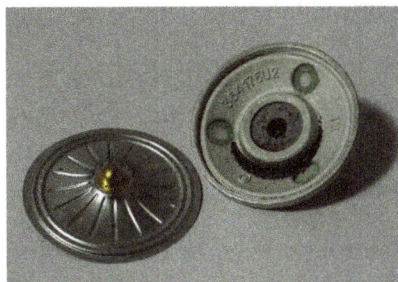

A disassembled Ericsson carbon microphone with carbon particles visible

Before the proliferation of vacuum tube amplifiers in the 1920s, carbon microphones were the only practical means of obtaining high-level audio signals. They were widely used in telephone systems until the 1980s, while other applications used different microphone designs much earlier. Their low cost, inherently high output and frequency response characteristic were well suited for telephony. For plain old telephone service (POTS), carbon-microphone based telephones can still be used without modification. Carbon microphones, usually modified telephone transmitters, were widely used in early AM radio broadcasting systems, but their limited frequency response, as well as a fairly high noise level, led to their abandonment in those applications by the late 1920s. They continued to be widely used for low-end public address, and military and amateur radio applications for some decades afterward.

History

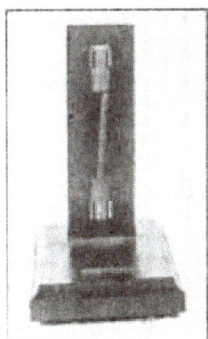

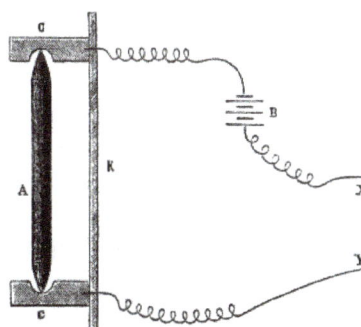

Hughes first carbon microphone, consisting of a carbon bar suspended loosely between two metal contacts with current from a battery passing through. The sound waves vibrated the bar, varying the resistance of the carbon contact points, varying the current.

The first microphone that enabled proper voice telephony was the (loose-contact) carbon microphone (then called transmitter). This was independently developed by David Edward Hughes in England and Emile Berliner and Thomas Edison in the US. Although Edison was awarded the first patent in mid-1877, Hughes had demonstrated his working device in front of many witnesses some years earlier, and most historians credit him with its invention.

Hughes' device used loosely packed carbon granules - the varying pressure exerted on the granules by the diaphragm from the acoustic waves caused the resistance of the carbon to vary proportionally, allowing a relatively accurate electrical reproduction of the sound signal. Hughes also coined the word microphone. He demonstrated his apparatus to the Royal Society by magnifying the sound of insects scratching through a sound box. Contrary to Edison, Hughes decided not to take out a patent; instead, he made his invention a gift to the world.

In America, Edison and Berliner fought a long legal battle over the patent rights. Ultimately a federal court awarded Edison full rights to the invention, stating "Edison preceded Berliner in the transmission of speech...The use of carbon in a transmitter is, beyond controversy, the invention of Edison" and the Berliner patent was ruled invalid.

The carbon microphone is the direct prototype of today's microphones and was critical in the development of telephony, broadcasting and the recording industries. Later, carbon granules were used between carbon buttons. Carbon microphones were widely used in telephones from 1890 until the 1980s.

Use as Amplifier

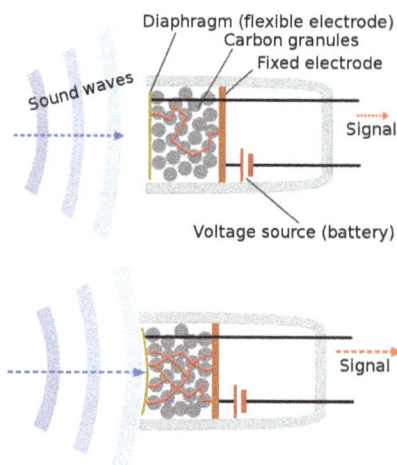

Operation of carbon microphone. When a sound wave presses on the conducting diaphragm, the granules of carbon are pressed together and decrease their electrical resistance.

Carbon microphones can be used as amplifiers. This capability was used in early telephone repeaters, making long distance phone calls possible in the era before vacuum tube amplifiers. In these repeaters, a magnetic telephone receiver (an electrical-to-mechanical transducer) was mechanically coupled to a carbon microphone. Because a carbon microphone works by varying a current passed through it, instead of generating a signal voltage as with most other microphone types, this arrangement could be used to boost weak signals and send them down the line. These amplifiers were mostly abandoned with the development of vacuum tubes, which offered higher gain and better sound quality. Even after vacuum tubes were in common use, carbon amplifiers continued to be used during the 1930s in portable audio equipment such as hearing aids. The Western Electric 65A carbon amplifier was 1.2" in diameter and 0.4" high and weighed less than 1.4 ounces. Such carbon amplifiers did not require the heavy bulky batteries and power supplies used by vacuum tube amplifiers. By the 1950s, carbon amplifiers for hearing aids had been replaced by miniature vacuum tubes (only to be shortly replaced by transistors). However, carbon amplifiers are still being produced and sold.

An illustration of the amplification provided by carbon microphones was the oscillation caused by feedback, that resulted in an audible squeal from the old "candlestick telephone" if its earphone was placed near the carbon microphone.

Early Radio Applications

Early AM radio transmitters relied on carbon microphones for voice modulation of the radio signal. In the first long-distance audio transmissions by Reginald Fessenden in 1906, a continuous wave from an Alexanderson alternator was fed directly to the transmitting antenna through a water-cooled carbon microphone. Later systems using vacuum tube oscillators often used the output from a carbon microphone to modulate the grid bias of the oscillator or output tube to achieve modulation.

Current Usage

Apart from legacy telephone installations in Third World countries, carbon microphones are still used today in certain niche applications in the developed world. An example is the Shure 104c, which is still in demand because of its wide compatibility with existing equipment.

The principal advantage of carbon microphones over other microphone designs is that they can produce high-level audio signals from very low DC voltages, without needing any form of additional amplification or batteries. This is particularly useful in remote locations served by very long telephone lines, where the electrical resistance of the wires can lead to severe DC voltage drop. Most all-electronic telephones need at least three volts DC to work, and so will often become useless in such situations, whereas carbon transmitter telephones will continue to work down to a fraction of a volt. Even where they do work, electronic telephones also suffer from the so-called "cliff effect", whereby they abruptly stop working when the line voltage falls below the critical level. In particular, this means that one telephone on a "party line" may tend to "hog" all the line current, cutting the others off. With carbon microphones, all receivers on the same line will still operate, albeit with reduced output.

Carbon microphones are also widely used in safety-critical applications such as mining and chemical manufacturing, where higher line voltages cannot be used, due to the risk of sparking and consequent explosions. Carbon-based telephone systems are also resistant to damage from high-voltage transients, such as those produced by lightning strikes, and electromagnetic pulses of the type generated by nuclear explosions, and so are still maintained as backup communication systems in critical military installations.

Class-D Amplifier

A class-D amplifier or switching amplifier is an electronic amplifier in which the amplifying devices (transistors, usually MOSFETs) operate as electronic switches, and not as linear gain devices as in other amplifiers. The signal to be amplified is a train of constant amplitude pulses, so the active devices switch rapidly back and forth between a fully conductive and nonconductive state. The analog signal to be amplified is converted to a series of pulses by pulse width modulation, pulse density modulation or other methods before being applied to the amplifier. After amplification, the output pulse train can be converted back to an analog signal by passing through a passive low pass filter consisting of inductors and capacitors.

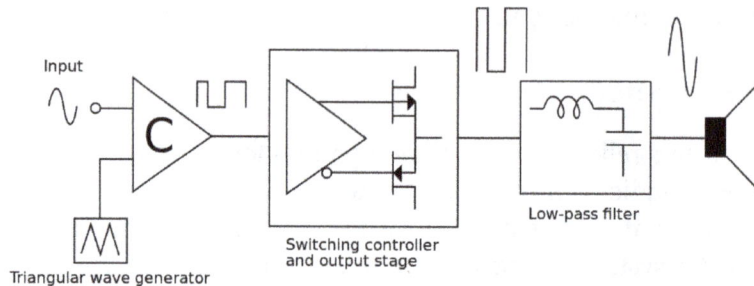

Block diagram of a basic switching or PWM (class-D) amplifier Note: For Clarity, Signal Periods are not shown to scale

Basic Operation

Class-D amplifiers work by generating a train of square pulses of fixed amplitude but varying width and separation, the low-frequency portion of whose frequency spectrum is essentially the signal to be amplified. The high-frequency portion serves no purpose other than to create a two level waveform. Because it has only two levels, it can be amplified by simple switching. The output of such a switch is an identical train of square pulses, except with greater amplitude. Such amplification results in a waveform with the same frequency spectrum, but with every frequency uniformly magnified in amplitude.

A passive low-pass filter removes the unwanted high-frequency components, i.e., smooths the pulses out and recovers the desired low-frequency signal. To maintain high efficiency, the filter is made with purely reactive components (inductors and capacitors), which store the excess energy until it is needed instead of converting some of it into heat. The switching frequency is typically chosen to be ten or more times the highest frequency of interest in the input signal. This eases the requirements placed on the output filter. In cost sensitive applications the output filter is sometimes omitted. The circuit then relies on the inductance of the loudspeaker to keep the HF component from heating up the voice coil. It will also need to implement a form of three-level (class-BD) modulation which reduces HF output, particularly when no signal is present.

The structure of a class-D power stage is essentially identical to that of a synchronously rectified buck converter (a type of non-isolated switched-mode power supply (SMPS)). Whereas buck converters usually function as voltage regulators, delivering a constant DC voltage into a variable load and can only source current (one-quadrant operation), a class-D amplifier delivers a constantly changing voltage into a fixed load, where current and voltage can independently change sign (four-quadrant operation). A switching amplifier must not be confused with linear amplifiers that use an SMPS as their source of DC power. A switching amplifier may use any type of power supply (e.g., a car battery or an internal SMPS), but the defining characteristic is that the amplification process itself operates by switching.

Theoretical power efficiency of class-D amplifiers is 100%. That is to say, all of the power supplied to it is delivered to the load, none is turned to heat. This is because an ideal switch in its on state would conduct all the current but have no voltage loss across it, hence no heat would be dissipated. And when it is off, it would have the full supply voltage across it but no leak current flowing through it, and again no heat would be dissipated. Real-world power MOSFETs are not ideal switches, but practical efficiencies well over 90% are common. By contrast, linear AB-class amplifiers are always operated with both current flowing through and voltage standing across the power devices. An ideal class-B amplifier has a theoretical maximum efficiency of 78%. Class A amplifiers (purely linear, with the devices always "on") have a theoretical maximum efficiency of 50% and some versions have efficiencies below 20%.

Terminology

The term "class D" is sometimes misunderstood as meaning a "digital" amplifier. While some class-D amps may indeed be controlled by digital circuits or include digital signal processing devices, the power stage deals with voltage and current as a function of non-quantized time. The smallest amount of noise, timing uncertainty, voltage ripple or any other non-ideality immediately results in an irreversible change of the output signal. The same errors in a digital system will only lead to incorrect results when they become so large that a signal representing a digit is distorted beyond recognition. Up to that point, non-idealities have no impact on the transmitted signal. Generally, digital signals are quantized in both amplitude and wavelength, while analog signals are quantized in one (e.g. PWM) or (usually) neither quantity.

Signal Modulation

The 2-level waveform is derived using pulse-width modulation (PWM), pulse density modulation (sometimes referred to as pulse frequency modulation), sliding mode control (more commonly called "self-oscillating modulation" in the trade.) or discrete-time forms of modulation such as delta-sigma modulation.

The most basic way of creating the PWM signal is to use a high speed comparator ("C" in the block-diagram above) that compares a high frequency triangular wave with the audio input. This generates a series of pulses of which the duty cycle is directly proportional with the instantaneous value of the audio signal. The comparator then drives a MOS gate driver which in turn drives a pair of high-power switches (usually MOSFETs). This produces an amplified replica of the comparator's PWM signal. The output filter removes the high-frequency switching components of the PWM signal and recovers the audio information that the speaker can use.

DSP-based amplifiers which generate a PWM signal directly from a digital audio signal (e. g. SPDIF) either use a counter to time the pulse length or implement a digital equivalent of a triangle-based modulator. In either case, the time resolution afforded by practical clock frequencies is only a few hundredths of a switching period, which is not enough to ensure low noise. In effect, the pulse length gets quantized, resulting in quantization distortion. In both cases, negative feedback is applied inside the digital domain, forming a noise shaper which has lower noise in the audible frequency range.

Design Challenges

Switching Speed

Two significant design challenges for MOSFET driver circuits in class-D amplifiers are keeping dead times and linear mode operation as short as possible. "Dead time" is the period during a switching transition when both output MOSFETs are driven into Cut-Off Mode and both are "off". Dead times need to be as short as possible to maintain an accurate low-distortion output signal, but dead times that are too short cause the MOSFET that is switching on to start conducting before the MOSFET that is switching off has stopped conducting. The MOSFETs effectively short the output power supply through themselves in a condition known as "shoot-through". Meanwhile, the MOSFET drivers also need to drive the MOSFETs between switching states as fast as possible to minimize the amount of time a MOSFET is in Linear Mode—the state between Cut-Off Mode and Saturation Mode where the MOSFET is neither fully on nor fully off and conducts current with a significant resistance, creating significant heat. Driver failures that allow shoot-through and/or too much linear mode operation result in excessive losses and sometimes catastrophic failure of the MOSFETs.

Electromagnetic Interference

The switching power stage generates both high dV/dt and dI/dt, which give rise to radiated emission whenever any part of the circuit is large enough to act as an antenna. In practice, this means the connecting wires and cables will be the most efficient radiators so most effort should go into preventing high-frequency signals reaching those:

- Avoid capacitive coupling from switching signals into the wiring.

- Avoid inductive coupling from various current loops in the power stage into the wiring.

- Use one unbroken ground plane and group all connectors together, in order to have a common RF reference for decoupling capacitors

- Include the equivalent series inductance of filter capacitors and the parasitic capacitance of filter inductors in the circuit model before selecting components.

- Wherever ringing is encountered, locate the inductive and capacitive parts of the resonant circuit that causes it, and use parallel RC or series RL snubbers to reduce the Q of the resonance.

- Do not make the MOSFETs switch any faster than needed to fulfil efficiency or distortion requirements. Distortion is more easily reduced using negative feedback than by speeding up switching.

Power Supply Design

Class-D amplifiers place an additional requirement on their power supply, namely that it be able to sink energy returning from the load. Reactive (capacitive or inductive) loads store energy during part of a cycle and release some of this energy back later. Linear amplifiers will dissipate this energy away, class-D amplifiers return it to the power supply which should somehow be able to store it. In addition, half-bridge class D amps transfer energy from one supply rail (e.g. the positive rail) to the other (e.g. the negative) depending on the sign of the output current. This happens regardless of whether the load is resistive or not. The supply should either have enough capacitive storage on both rails, or be able to transfer this energy back.

Error Control

The actual output of the amplifier is not just dependent on the content of the modulated PWM signal. The power supply voltage directly amplitude-modulates the output voltage, dead time errors make the output impedance non-linear and the output filter has a strongly load-dependent frequency response. An effective way to combat errors, regardless of their source, is negative feedback. A feedback loop including the output stage can be made using a simple integrator. To include the output filter, a PID controller is used, sometimes with additional integrating terms. The need to feed the actual output signal back into the modulator makes the direct generation of PWM from a SPDIF source unattractive. Mitigating the same issues in an amplifier without feedback requires addressing each separately at the source. Power supply modulation can be partially canceled by measuring the supply voltage to adjust signal gain before calculating the PWM and distortion can be reduced by switching faster. The output impedance cannot be controlled other than through feedback.

Advantages

The major advantage of a class-D amplifier is that it can be more efficient than an analog amplifier, with less power dissipated as heat in the active devices. Given that large heat sinks are not required, Class-D amplifiers are much lighter weight than analog amplifiers, an important consideration with portable sound reinforcement system equipment and bass amplifiers. Output stages such as those used in pulse generators are examples of class-D amplifiers. However, the term mostly applies to power amplifiers intended to reproduce audio signals with a bandwidth well below the switching frequency.

Despite the complexity involved, a properly designed class-D amplifier offers the following benefits:

- Reduced power waste as heat dissipation and hence:

- Reduction in cost, size and weight of the amplifier due to smaller (or no) heat sinks, and compact circuitry,

- Very high power conversion efficiency, usually better than 90% above one quarter of the amplifier's maximum power, and around 50% at low power levels.

Boss Audio mono amp. The output stage is top left, the output chokes are the two yellow toroids underneath.

Uses

- Home theatre in a box systems. These economical home cinema systems are almost universally equipped with class-D amplifiers. On account of modest performance requirements and straightforward design, direct conversion from digital audio to PWM without feedback is most common.

- Mobile phones. The internal loudspeaker is driven by up to 1 W. Class D is used to preserve battery lifetime.

- Hearing aids. The miniature loudspeaker (known as the receiver) is directly driven by a class-D amplifier to maximise battery life and can provide saturation levels of 130 dB SPL or more.

- Powered speakers

- High-end audio is generally conservative with regards to adopting new technologies but class-D amplifiers have made an appearance

- Active subwoofers

- Sound Reinforcement and Live Sound. For very high power amplification the powerloss of AB amplifiers are unacceptable. Amps with several kilowatts of output power are available as class-D. The Crest Audio CD3000, for example, is a class-D power amplifier that is rated at 1500 W per channel, yet it weighs only 21 kg (46 lb). Similarly, the Powersoft K20 is a class-D power amplifier that is rated at 9000 W per 2-Ohm channel, yet it weighs only 12 kg (26.5 lb).

- Bass amplifiers. Again, an area where portability is important. Example: Yamaha BBT500H bass amplifier which is rated at 500 W, and yet it weighs less than 5 kg (11 lb). The Promethean P500H by Ibanez is also capable of delivering 500 W into a 4 Ohm load, and weighs only 2.9 kg (6.4 lb). Gallien Krueger MB500 and Eden WTX500, also rated at 500 W weighs no more than 2 kg (4.4 lb).

Headphone Amplifier

A Little Dot Mk III tube headphone amplifier.

A headphone amplifier is a low-powered audio amplifier designed particularly to drive headphones worn on or in the ears, instead of loudspeakers in speaker enclosures. Most commonly, headphone amplifiers are found embedded in electronic devices that have a headphone jack, such as integrated amplifiers, portable music players (e.g., iPods), and televisions. However, standalone units are used, especially in audiophile markets and in professional audio applications, such as music studios. Headphone amplifiers are available in consumer-grade models used by hi-fi enthusiasts and audiophiles and professional audio models, which are used in recording studios.

Consumer Models

The electronics of a Musical Fidelity X-Cans, using two 6922 vacuum tubes (electronic valves)

Consumer headphone amplifiers are commercially available separate devices, sold to a niche audiophile market of hi-fi enthusiasts. These devices allow for higher possible volumes and superior current capacity compared to the smaller, less expensive headphone amplifiers that are used in most audio players. In the case of the extremely high-end electrostatic headphones, such as the Stax SR-007, a specialized electrostatic headphone amplifier or transformer step-up box and power amplifier is required to use the headphones, as only a dedicated electrostatic headphone amplifier or transformer can provide the voltage levels necessary to drive the headphones. Most headphone amplifiers provide power between 10 mW and 2 W depending on the specific headphone being used and the design of the amplifier. Certain high power designs can provide up to 6W of power into low impedance loads, although the benefit of such power output with headphones is unclear, as the few orthodynamic headphones that have sufficiently low sensitivities to function with such power levels will reach dangerously high volume levels with such amplifiers.

Effectively, a headphone amplifier is a small power amplifier that can be connected to a standard headphone jack or the line output of an audio source. Electrically, a headphone amplifier can be thought of as an amplifier that presents a very high input impedance (ideally infinite) and presents a lower output impedance (ideally zero) and larger range of output voltages (ideally infinite). This allows headphones of a low sensitivity to be driven louder as a result of the extra voltage provided by the amplifier. There are potential fidelity gains if headphones are driven with lower distortion than using a headphone amplifier integrated into a general purpose audio product. In practice, this most often occurs when using low impedance headphones with consumer electronics with insufficiently low output impedance.

Volume

A hybrid headphone amplifier with the chassis cover removed.

Most headphone amplifiers support a higher voltage output and therefore higher power (volume) levels. Whereas most portable electronics run off of a 1.8, 2.5 or 3.3 Vpp supply, many headphone amplifiers run off of 10, 18 or 24 Vpp volt supplies, allowing 5-20 dB higher volume. If a pair of headphones is too quiet, adding an amplifier that can output higher voltage/power will increase its volume.

Output Impedance

Many headphone amplifiers have an output impedance in the range of 0.5 - 50 Ohms. The 1996 IEC 61938 standard recommended an output impedance of 120 Ohms, but in practice this is rarely used and not recommended with modern headphones. High output impedance can result in frequency response fluctuations, due to varying load impedance at different frequencies. In 2008 Stereophile Magazine published an article that showed that a 120-Ohm output impedance could cause a 5-dB error in frequency response with certain types of headphones. However, the author of the article also states: "The ramifications for subjective assessment of headphones are more troublesome because it is usually unclear what assumptions the manufacturer has made regarding source impedance."

More importantly, low output impedance can reduce distortion by improving the control that the source has over the transducer. This is often expressed as damping factor, with higher damping factors greatly reducing distortion. One company shows a 45 dB improvement in THD+N at 30 Hz for their low-impedance amplifier compared to a 30-ohm amplifier. For example, a 32 Ω headphone driven by a headphone amp with a <1 Ω output impedance would have a damping factor of >32, whereas the same headphone driven with an iPod touch 3G (7 Ω output impedance) would have a damping factor of just 4.6. If the 120 ohms recommendation is applied, the damping factor would be an unacceptably low 0.26 and consequently distortion would be significantly higher. Conversely, the same iPod touch driving a pair of 120 ohm headphones would have a respectable damping factor of 17.1, and would most likely not benefit from the addition of a lower impedance headphone amplifier.

In addition to output impedance, other specifications are relevant to choosing a headphone amplifier — THD, frequency response, IMD, output power, minimum load impedance, and other measurements are also significant. However, most of these will be improved by lowering output impedance and hence improving damping factor.

DIY Approach

For those who are electronically inclined, the low-power and fairly simple nature of the headphone amplifier has made it a popular DIY project. There are many designs for headphone amplifiers posted on the Internet varying considerably in complexity and cost. A key example is the simple opamp-based CMoy design, one of the most popular headphone amplifier designs available. The simplicity of the CMoy makes it an easy build, while it can be made small enough to fit inside a tin of breath mints (including batteries). On the other hand, it is often built using op-amps that are not designed to drive loads as low as headphones, leading to poor performance and audible differences between op-amps that would not exist in a good design.

Crossfeed and other Audio Processing

Crossfeeding blends the left and right stereo channels slightly, reducing the extreme channel separation which is characteristic of headphone listening in older stereo recordings, and is known to cause headaches in a small fraction of listeners. Crossfeed also improves the soundstage characteristics and makes the music sound more natural, as if one was listening to a pair of speakers. While some swear by crossfeed, many prefer amplifiers without it. The introduction of digital signal processing (DSP) technology led a number of manufacturers to introduce amplifiers with 'headphone virtualization' features. In principle, the DSP chips allow the two-driver headphone to simulate a full Dolby 5.1 (or more) surround system. When the sounds from the two headphone drivers mix, they create the phase difference the brain uses to locate the source of a sound. Through most headphones, because the right and left channels do not combine as they do with crossfeed, the illusion of sound directionality is created.

Professional Audio Models

Distribution headphone amp - front panel

Distribution headphone amp - back panel

In pro-audio terminology, a headphone amplifier is a device that allows multiple headsets to be connected to one or more audio sources (typically balanced audio sources) at the same time to monitor sounds during a recording session, either singing or playing from the "live room" or recorded tracks. Headphone amps enable singers and musicians to be able to hear other musicians who are playing in isolation booths. They also enable audio engineers and record producers to monitor a live performance or live tracking.

Headphone amps with sub-mixing capabilities allow the listener to adjust, mix and monitor audio signals coming from multiple sources at the same time. This kind of headphone amp is often utilized during recording sessions to sub-mix playback of individual stem-mixes or instruments coming from a mixing board or a playback device. In many cases the listeners have their own sets of controls allowing them to adjust various aspects of the mix and individual and global parameters such as channel level, global loudness, bass and treble.

Distribution headphone amplifiers are specialized headphone amps allowing a single signal to be fed to multiple headsets or multiple groups of multiple headsets at the same time. Many distribution headphone amps, like the one shown here, can be cascaded by connecting the audio input of one of the amps to the cascading output, marked "THRU", of another amp.

There are also various other combinations of pro-audio headphone amps with simultaneous sub-mixing and distribution capabilities.

Class-T Amplifier

Two Tripath chipset Class T stereo amplifier modules. TA2024 6+6W to the left, TA2020 20+20W to the right

Class T was a registered trademark for a switching (class-D) audio amplifier, used for Tripath's amplifier technologies (patent filed on Jun 20, 1996). Similar designs have now been widely adopted by different manufacturers.

The covered products use a class-D amplifiers combined with proprietary techniques to control the pulse width modulation to produce what is claimed to be better performance than other class-D amplifier designs. Among the publicly disclosed differences is real time control of the switching frequency depending on the input signal and amplified output. One of the amplifiers, the TA2020, was named one of the twenty-five chips that 'shook the world" by the IEEE Spectrum magazine.

The control signals in Class T amplifiers may be computed using digital signal processing or fully analog techniques. Currently available implementations use a loop similar to a higher order Delta-Sigma ($\Delta\Sigma$) (or sigma-delta) modulator, with an internal digital clock to control the sample comparator. The two key aspects of this topology are that (1), feedback is taken directly from the switching node rather than the filtered output, and (2), the higher order loop provides much higher loop gain at high audio frequencies than would be possible in a conventional single pole amplifier.

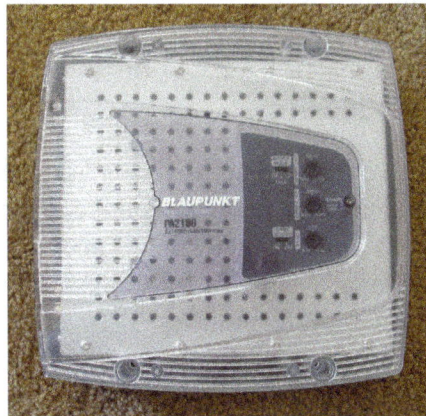

Blaupunkt PA2150 T-Amp, "Powered by Tripath"

Financial difficulties caused Tripath to file for Chapter 11 bankruptcy protection on 8 February 2007. Tripath's stock and intellectual property were purchased later that year by Cirrus Logic.

Products and Applications

Tripath used to sell the amplifiers as chips, or as chipsets, to be integrated into products by other companies in several countries. For example:

- Sony, Panasonic and Blaupunkt use them in several car stereos and integrated home cinema systems

- Apple Computer used them in their Power Mac G4 Cube, Power Mac G4 (Digital audio), eMac and iMac (Flat Panel) computers

- Audio Research, an audio electronics company, formerly an exclusive tube circuit specialist, produced a Tripath-based audiophile amplifier.

- Sonneteer, a British hifi company, analogue amplifier and technology specialists, manufactured the Bronte amplifier (1999), based around a Tripath class-T module.

- Bel Canto Design used them in their eVo range of amplifiers.

- Sonneteer, a British hifi company, analogue amplifier and technology specialists, manufactured under Sub Brand Bard Audio, the BardThree amplifier (2005), based around a Tripath class-T chip and is/was known as the Plug Top or Rug Rat amp.

Valve Audio Amplifier

A valve audio amplifier (UK) or vacuum tube audio amplifier (United States) is a valve amplifier used for sound reinforcement, sound recording and reproduction.

Until the invention of solid state devices such as the transistor, all electronic amplification was produced by valve (tube) amplifiers. While solid-state devices prevail in most audio amplifiers today, valve audio amplifiers are still used where their audible characteristics are considered pleasing, for example in music performance or music reproduction.

Instrument and Vocal Amplification

Valve amplifiers for guitars (and to a lesser degree vocals and other applications) have different purposes from those of hi-fi amplifiers. The purpose is not necessarily to reproduce sound as accurately as possible, but rather to fulfill the musician's concept of what the sound should be. For example, distortion is almost universally considered undesirable in hi-fi amplifiers but may be considered a desirable characteristic in performance.

Small signal circuits are often deliberately designed to have very high gain, driving the signal far outside the linear range of the tube circuit, to deliberately generate large amounts of harmonic distortion. The distortion and overdrive characteristics of valves are quite different from transistors (not least the amount of voltage headroom available in a typical circuit) and this results in a distinctive sound. Amplifiers for such performance applications typically retain tone and filter circuits that have largely disappeared from modern hi-fi products. Amplifiers for guitars in particular may also include a number of "effects" functions.

The Origins of Electric Guitar Amplification

The electric guitar originates from Rickenbacker in the 1930s but its modern form was popularised by Fender and Gibson (notably the Fender Telecaster (1951) & Stratocaster (1954) and Gibson Les Paul (1952) during the 1950s. The earliest guitar amplifiers were probably audio amplifiers made for other purposes and pressed into service, but the electric guitar and its amplification quickly developed a life of its own, supported by specialist manufacturers.

Rear view of a valve combo guitar amplifier. Visible are two glass 6L6 output tubes, six smaller 12AX7 preamp tubes in their metal tube retainers and both the power transformer and the output transformer.

Guitar amplifiers are often designed so they can, when desired by the guitarist, distort and create a tone rich in harmonics and overtones. The characteristics of the tube and the circuit directly influence the nature of the sound produced. Even the power supply can influence the tonal shape, with relatively undersized power supply capacitors producing a characteristic "sag" at instants of peak output and power draw, and subsequent recovery, that is often considered musically engaging. In addition, guitarists may employ acoustic feedback, further modifying the resulting sound (noting that the feedback signal has a slight time lag relative to the original signal).

Guitar amplifiers are typically designed to withstand a lot of abuse both electrically and physically (since guitarists often travel to gigs, etc.) In large systems the amplifier is separate from the speaker enclosure(s), but in smaller systems it is often integrated, forming a so-called "combo". Since the amplifier is usually at the top of the combo, the tubes often hang upside down facing the body of the enclosure. They may be held in with clips.

Most modern valve guitar amplifiers use a class AB1 push-pull circuit with a pair of power pentodes or beam tetrodes, 6L6 or EL34 but occasionally KT88, 6550, or the lower-power EL84 in *Ultra-Linear* connection. The output stage is preceded by a voltage amplification stage (pentode or twin triode) and a phase-splitter (twin triode). Twin triodes with two identical sections in one envelope are used, usually the noval types 12AT7, 12AU7, or 12AX7 or equivalents, less usually the octal 6SN7.

Amplifiers for Sound Reproduction

Early Development

The earliest mass usage of valve audio amplifiers was for telephony. Valve amplifiers were critical in development of long-distance telephone circuits and submarine telephone cables. Radio applications followed soon after, where valves were used for both the audio (AF) and radio (RF) circuitry.

Among the first applications of sound recording and electronic replay around the 1920s was its use in many cinemas equipping for exhibiting the new 'talkies'. Cinema sound systems of this period were predominantly supplied by "Westrex", related to the Western Electric company, a telecoms supplier, who were also the makers of the 300B DHT tube that today is central to current production DH-SET audiophile amplification.

Amplifiers during this period typically used Directly Heated tubes in a Class A Single-Ended Triode circuit. Power output ranged from a few watts to perhaps 20 watts for an exceptionally powerful amplifier (modern semiconductor amplifiers produce much higher power). Today this type of circuit retains a niche following at the very extreme of audiophile hi-fi, where it is often referred as DH-SET.

Prior to WWII, almost all electronic amplifiers were triodes used without feedback. The inherent, albeit imperfect, linearity of tubes makes it possible to get acceptable distortion performance without correction. Amplitude distortion in a class A triode stage can be small if care is taken to prevent the anode current from becoming too small and ensuring that grid current does not flow at any point. In this case, distortion is largely relatively unobjectionable second harmonic, with percentage closely proportional to the output amplitude. Adding modest negative feedback improves linearity further. Pentodes of the same power dissipation are capable of higher output power than triodes, but distortion is higher and more objectionable.

The 1940s and 1950s

During the post-war period, widespread adoption of negative feedback in the push pull topology yielded greater power and linearity, notably following the publication in 1947 of the Williamson amplifier, which set the standard (and the dominant topology) for what was to follow.

Widespread adoption of push pull allowed smaller (and thus cheaper) transformers, combined with more power (typically 10 to 20 watts) to handle peaks. The high fidelity industry was born.

Other developments included (among others):

- The introduction of the "Point One" series of amplifiers (in 1945) by LEAK in the UK, which first set a performance standard of 0.1% THD

- The Ultra-Linear output stage (a tapped push-pull output transformer providing power intermediate between triodes and pentodes, at lower distortion than either) was originated by Alan Blumlein in 1937 in the UK, but popularised following publication of a paper by David Hafler and Keroes in the USA in 1951, and became the dominant topology during the post war recovery of consumer products

- Manufacturers bringing high-quality domestic hi-fi to a steadily widening audience, eventually leading to Dynaco selling over 300,000 ST-70's, making it the world's most popular hi-fi amplifier of any type to date.

Tube Hi-fi in the 1960s

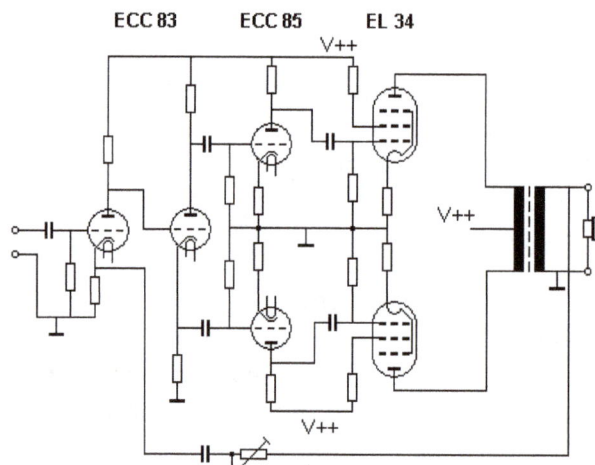

Conceptual diagram of a poweramplifier with a split-load phase inverter and push-pull EL34 pentodes endstage

Valve amplification peaked as the mainstream technology during the 1960s and 70s, with device and circuits being highly developed. There have been only minor refinements since then.

The last generation of power tubes, typified by KT66, EL34 and KT88, represent the pinnacle of the technology and of production quality. Valve amplifiers produced since that time usually use one of these tubes, which have remained in continuous production (apart from KT66) ever since. Output power was typically 20 watts, exceptionally 35W.

Small signal valves overwhelmingly changed from octal base tubes, notably the audio tube of choice, the 6SN7 family, to the smaller and cheaper noval base ECC81, ECC82, ECC83 (UK, in the US known as 12AX7,12AT7, etc.). The lower-power noval base EL84 power pentode was widely used in less expensive 10-watt ultralinear power amplifiers, still of high fidelity.

Commercial tube manufacturers developed designs based on their own products, most notably the Mullard 5-10 circuit. This design and the Williamson were widely implemented and imitated, with or without crediting the originator.

Automobile Amplifiers

Valve radios and amplifiers were used in automobiles until they were displaced by transistorized radios. Transistors had the major advantage of working off the voltage provided by a car battery. Early radios required a power unit to convert the battery voltage to a value high enough for the valves. Later radios used special valves that were designed to operate directly from a 12 volt supply. These later radios were hybrid designs which used transistors only for the audio output stages because a 12 volt power amplifier valve was not practical. For this and other applications, transistors are smaller, cheaper, more durable, use less power, run cooler, and do not need to warm up.

Some enthusiasts prefer "tube amps", so a small number of valve car stereos are still made. Manufacturers include Milbert Amplifiers, Blade, Manley, and Sear Sound. Some are hybrid designs with transistors and valves.

Valve Preamplifiers

Due to the very poor technical performance of early gramophones, the lack of standardised equalisations, poor components and accessories (including loudspeakers), preamplifiers historically contained extensive and very flexible equalization and tone and filter circuits designed to adjust the frequency response of the amplifier and so the sound produced by the system.

Valve preamplifiers use triodes or low-noise pentodes (EF86). Mains hum from the heater filaments is a potential problem in low-level valve stages. Modern amplifiers invariably run from the mains; as there is little need to minimise costs in expensive valve amplifiers, the heater supply is often rectified and even regulated to reduce hum to an absolute minimum.

A representative valve preamp from the 1950s is the Leak 'varislope' series of preamps, which included a switchable rumble filter, a switchable scratch filter with selectable slopes and corner frequency, continuously variable treble and bass tone controls and a selection of 4 different gramophone equalisations (RIAA, ortho, RCA, 78).

Valve Sound

Amplifiers from and prior to this period often have a distinctive sound that today is still widely referred to as "valve sound" and "warm". This tone is not strictly caused by the use of valves rather than transistors; it is merely a sound that was originally associated with amplifiers built using valves simply because that is what was available at the time. The origins of that particular sound are in part due to:

- the typical circuit designs of the time (class A or AB1), combined with

  o simple circuits with relatively little feedback, producing mainly low-harmonic distortion (feedback that could be applied was limited by phase shift in the output transformer);

  o low damping factor (High Z out) output stages;

  o the use of very large amounts of feedback in signal transformerless semiconductor circuits, which leads to distortion that is small but has a much larger proportion of high harmonics than in valve circuits.

Factors relevant to valve equipment not designed to high-fidelity standards:

- dedicated guitar amplifiers had frequency response and distortion suited to their purpose,

- under-dimensioned and unregulated power supplies in amplifiers where power drawn varied significantly with instantaneous power output (class A not affected, classes AB1, AB2, B, increasingly so),

- poor quality output transformers in budget equipment.

Notable Historic Designs

Quad II power amplifier

In addition to a range of commodity valve amplifiers, some amplifiers were made which are still highly regarded today. Among the best known are:

- LEAK TL/12

- Williamson amplifier

- Mullard 5-10

- Quad II

- Dynaco Mark III and Stereo 70

- McIntosh MC275

- Marantz 8B and 9

Valve Audio Amplifier Technical Information

Various basic circuits have been used in designs and in various approaches to construction.

References

- Bode, H. W. (July 1940). "Relations Between Attenuation and Phase in Feedback Amplifier Design". Bell Labs Technical Journal. AT&T. 19 (3): 421–454. doi:10.1002/j.1538-7305.1940.tb00839.x

- Glisson, Tildon H. (2011). Introduction to Circuit Analysis and Design. Springer Science and Business Media. ISBN 9048194431

- Godfrey, Donald G. (1998). "Audion". Historical Dictionary of American Radio. Greenwood Publishing Group. p. 28. Retrieved January 7, 2013

- Robert S. Symons (1998). "Tubes: Still vital after all these years". IEEE Spectrum. 35 (4): 52–63. doi:10.1109/6.666962

- "Circuit Design Modifications for Minimizing Transient Intermodulation Distortion in Audio Amplifiers", Matti Otala, Journal of Audio Engineering Society, Vol 20 # 5, June 1972

- Patronis, Gene (1987). "Amplifiers". In Glen Ballou. Handbook for Sound Engineers: The New Audio Cyclopedia. Howard W. Sams & Co. p. 493. ISBN 0-672-21983-2

- McNicol, Donald (November 1, 1917). "The Audion Tribe". Telegraph and Telephone Age. New York: J. B. Taltavall. 21: 493. Retrieved May 12, 2017

- "Psychoacoustic Detection Threshold of Transient Intermodulation Distortion", Petri-Larmi, M.; Otala, M.; Lammasniemi, J. Journal of Audio Engineering Society, Vol 28 # 3, March 198

- Nebeker, Frederik (2009). Dawn of the Electronic Age: Electrical Technologies in the Shaping of the Modern World, 1914 to 1945. John Wiley and Sons. pp. 9–10, 15. ISBN 0470409746

- "David Edward Hughes: Concertinist and Inventor" (PDF). Archived from the original (PDF) on 2013-12-31. Retrieved 2012-12-17

- Roy, Apratim; Rashid, S. M. S. (5 June 2012). "A power efficient bandwidth regulation technique for a low-noise high-gain RF wideband amplifier". Central European Journal of Engineering. 2 (3): 383–391. Bibcode:2012CEJE....2..383R. doi:10.2478/s13531-012-0009-1

- Robert Boylestad and Louis Nashelsky (1996). Electronic Devices and Circuit Theory, 7th Edition. Prentice Hall College Division. ISBN 978-0-13-375734-7

- "Archived copy". Archived from the original on 2011-04-24. Retrieved 2011-01-16. Cyrus Audio: Product Archive: Cyrus One

- Otala, M. (1970). "Transient distortion in transistorized audio power amplifiers". IEEE Transactions on Audio and Electroacoustics. 18 (3): 234. doi:10.1109/TAU.1970.1162117

- Agarwal, Anant; Lang, Jeffrey (2005). Foundations of Analog and Digital Electronic Circuits. Morgan Kaufmann. p. 331. ISBN 008050681X

Permissions

We would like to thank the editorial team for lending their expertise to make the book truly unique. They have played a crucial role in the development of this book. Without their invaluable contributions this book wouldn't have been possible. They have made vital efforts to compile up to date information on the varied aspects of this subject to make this book a valuable addition to the collection of many professionals and students.

This book was conceptualized with the vision of imparting up-to-date and integrated information in this field. To ensure the same, a matchless editorial board was set up. Every individual on the board went through rigorous rounds of assessment to prove their worth. After which they invested a large part of their time researching and compiling the most relevant data for our readers.

The editorial board has been involved in producing this book since its inception. They have spent rigorous hours researching and exploring the diverse topics which have resulted in the successful publishing of this book. They have passed on their knowledge of decades through this book. To expedite this challenging task, the publisher supported the team at every step. A small team of assistant editors was also appointed to further simplify the editing procedure and attain best results for the readers.

Apart from the editorial board, the designing team has also invested a significant amount of their time in understanding the subject and creating the most relevant covers. They scrutinized every image to scout for the most suitable representation of the subject and create an appropriate cover for the book.

The publishing team has been an ardent support to the editorial, designing and production team. Their endless efforts to recruit the best for this project, has resulted in the accomplishment of this book. They are a veteran in the field of academics and their pool of knowledge is as vast as their experience in printing. Their expertise and guidance has proved useful at every step. Their uncompromising quality standards have made this book an exceptional effort. Their encouragement from time to time has been an inspiration for everyone.

The publisher and the editorial board hope that this book will prove to be a valuable piece of knowledge for students, practitioners and scholars across the globe.

Index

www.ingramcontent.com/pod-product-compliance
Lightning Source LLC
Chambersburg PA
CBHW082058190326
41458CB00010B/3524